Parasitology

LIFELINES

PARASITOLOGY

■ Jack Chernin
School of Biological Sciences,
University of Portsmouth, UK

London and New York

First published 2000 by Taylor & Francis
11 New Fetter Lane, London EC4P 4EE

Simultaneously published in the USA and Canada
by Taylor & Francis Inc,
29 West 35th Street, New York, NY 10001

Taylor & Francis is an imprint of the Taylor & Francis Group

Typeset in Perpetua and Helvetica by Graphicraft Limited, Hong Kong
Printed and bound in Great Britain by TJ International Ltd, Padstow, Cornwall

British Library Cataloguing in Publication Data
A catalogue record for this book is available from the British Library

Library of Congress Cataloging in Publication Data
Chernin, Jack.
Parasitology/Jack Chernin.
p. cm. — (Modules in life sciences)
Includes bibliographical references and index.
1. Medical parasitology. 2. Parasitology. I. Title. II. Series.

QR251.C52 2000 99-055177
616.9′6—dc21

ISBN 0-7484-0817-7

CONTENTS

SERIES EDITOR'S PREFACE

Teaching programmes in universities now are generally arranged in collections of discrete units. These go under various names such as units, modules, or courses. They usually stand alone as regards teaching and assessment but, as a set, comprise a programme of study. Usually around half of the units taken by undergraduates are compulsory and effectively define a 'core' curriculum for the final degree. The arrangement of teaching in this way has the advantage of flexibility. The range of options over and above the core curriculum allows the student to choose the best programme for her or his future.

The Lifelines series provides a selection of texts that can be used at the undergraduate level for subjects optional to the main programme of study. Each volume aims to cover the material at a depth suitable to a particular level or year of study, with an amount of material appropriate to around one quarter of the undergraduate year. The concentration on life science subjects in the Lifelines series reflects the fact that it is here where individual topics proliferate.

Suggestions for new subjects and comments on the present volumes in the series are always welcomed and should be addressed to the series editor.

John Wrigglesworth
London, March 2000

INTRODUCTION TO PARASITOLOGY

1

1.1 PARASITES AND PARASITISM

The word 'parasite' is derived from the Greek words *para* (meaning beside) and *sitos* (meaning food). Parasites can be described as living organisms that are associated with food for all or part of their life-cycle. The organism providing the food is generally called the host. A parasite has at least one host per life-cycle. If there is more than one host per life-cycle, the host in which sexual maturity occurs is referred to as the definitive host and the other host/s are known as intermediate hosts. The study of parasites invariably involves firstly the biology of the parasite and secondly the biology of the host — the parasite's environment. Hence the following comments can be made regarding parasitology:

- Parasitology can be considered to be a specialised branch of ecology.
- Parasitology is the study of organisms living within a specialised environment.

The problems related to survival for parasites are almost the opposite of those faced by free-living animals. Parasites are surrounded by (they live within or on) their food and do not need to spend energy to find food; whereas free living animals are continuously searching for food. Free living animals have fewer problems than parasites in reproduction and distribution. During the distributive phase of its life-cycle, the probability of a parasite making contact with a new host is relatively low.

The evolution of successful methods of invasion and escape is essential for the survival of a parasite. Parasites have evolved mechanisms to ensure distribution and making contact with a new host. Similar adaptive strategies that follow a basic 'parasitic' plan/design have evolved within different taxonomic groups. Parasites have to adapt to two basic environments:

- the micro-environment — the habitat within the host immediately surrounding the parasite;
- the macro-environment — the habitat of the host.

Like many free-living animals, parasites have become equipped to survive within a certain habitat. The majority of parasites occupy a specific or predetermined site

■ **BOX 1.1**

Parasitology as a discipline has to investigate all aspects of the following:

- The biology of the parasite.
- The variations in life-cycle of the parasites.
- Methods of invasion of the host.
- Migration and maturation within the host.
- The effect of the parasite upon the host.
- The host reaction and response to the parasite.
- Methods of escape from the host.
- Distribution of the parasite.

within their host to which they have become adapted. There are very few areas of the vertebrate body that have not been invaded by parasites adapted to survive in that particular microenvironment.

Parasites are often described according to which site they inhabit, such as:

- Intracellular or inside the membrane of host cells — intracellular parasites are more or less restricted to the size of these cells and are mostly microscopic.
- Extracellular or outside the cell, but within body fluids or in the ground substance or matrix of tissues and organs. They range is size from micro- to macroscopic.

One of the major problems that a parasite has to overcome is that the environment (the host) in which it lives also happens to be its source of food, but reacts against it. The presence of the parasite stimulates the host to try and destroy the parasite. In order to survive, the parasite has to try and avoid this reaction (see Box 1.1).

The majority of the parasites studied by modern parasitologists cause disease in humans or domestic animals. They are also known as parasites of economic importance. Nearly all of these parasites are invertebrates belonging to one of the following phyla or groups: the Protozoa, the Cestoda, the Trematoda, the Nematoda, the Acanthocephola or the Arthropoda. It is within these phyla that most parasites of economic importance are found.

It is almost certain that all parasites evolved from free-living forms. While certain species were evolving into larger aquatic and terrestrial creatures, others that had not evolved to the same extent chose to invade the larger forms. The successful invaders soon established their habitat within the larger organism. Over millions of years of evolutionary time both the parasite and its host adapted to (on the part of the parasite) and tolerated (on side of the host) each other. The younger the host–parasite relationship is in evolutionary terms the less tolerant the host will be of the invader. The host tries its utmost to exclude the invader for the simple reason that presence of parasites causes varying degrees of pathological damage to the host (see Box 1.2).

■ 1.1.1 METHODS OF INVASION AND ESCAPE

In order to survive as species the parasites must have some means of locating and invading new hosts. After development and maturation within the host the next generation must be able to escape from the host (see Fig. 1.1).

■ **BOX 1.2**

There are several types of parasitism each referring to the variation in the habitat and life of the particular parasites such as:

- Obligate parasites; organisms that for all or most of their life-cycle are parasitic.
- Temporary parasites; parasitic for limited periods for either feeding or reproduction.
- Facultative parasites; organisms that are not normally parasitic but can survive for a limited period when they accidentally find themselves within another organism.
- Adaptive parasites; those organisms that have the capacity to live both as free-living or parasitic organisms.
- Obligate parasites have at least one host during their life-history.
- If there is more than one host during the life-cycle, the host in which the parasite reaches sexual maturity is known as the definitive host.
- Hosts in which the larval stages are located are called the intermediate hosts.
- The stage of the parasite that invades the host is called the invasive or infective stage.

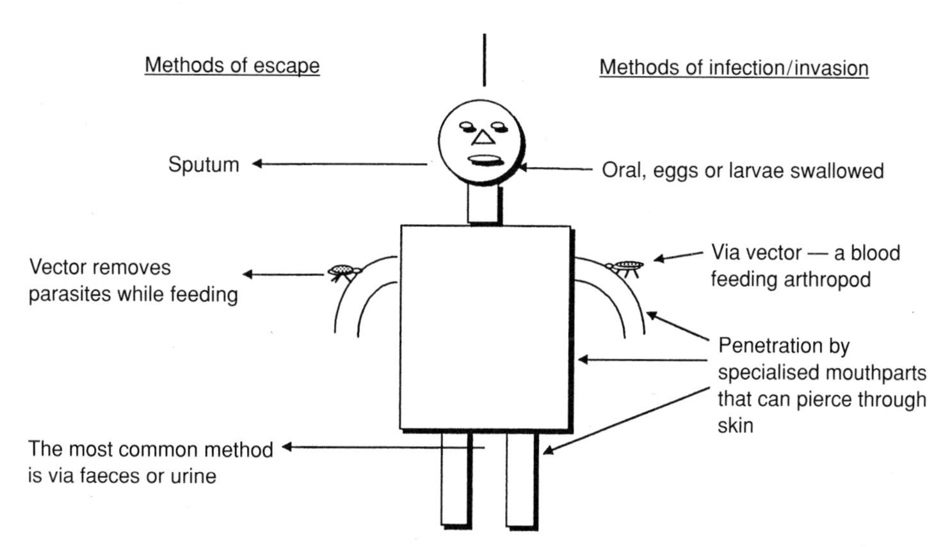

• **Figure 1.1** The survival of a parasite depends upon infecting a host and reproducing within the host environment. The next generation must be capable of escaping from the host to repeat the cycle. A parasite can enter and escape from a host via an orifice, or actively penetrate the host's outer covering or be infected and removed via a (arthropod) vector.

The following are the most frequently encountered methods of invasion:

- Oral: eggs or infectious larvae are swallowed via food or accidentally.
- Penetration by specialised larvae that, once they make contact with the host, can pierce through the skin or outer covering of the host.
- Via a vector: usually an arthropod that feeds upon the body fluids of a vertebrate transmits an infectious stage of the parasite. The vector can infect the host either by an inoculative method, ie injected via the arthropod mouth parts into the vertebrate, or by the contaminative method, ie the infectious stage is released from the vector while it is feeding, and the parasite then enters the host through the 'wound'.
- Contact: Infectious stages penetrate a host during copulation.

Methods of escape:

- The most common method of escape of eggs or larvae is via the faeces or urine. However escape can occur through any of the orifices.
- Via a vector, while feeding an arthropod could also passively consume a stage of the parasite.
- The parasite may actively break through the skin to escape.

1.1.2 LIFE-CYCLES

1.1.2.1 Life-cycles using a single host (Direct life-cycles)

Only one host is invaded during a single cycle of the parasite's life history. The parasite enters the host either as egg or an infectious larva. It grows, develops, matures and reproduces and then deposits eggs or larvae which escape to invade another host. Development usually proceeds only after the invasive stage begins to migrate to a selected or predetermined site. Reproduction within the host can be sexual, asexual, or both.

1.1.2.2 Life-cycles using two or more hosts (Indirect life-cycles)

At least two hosts are invaded during a single life-cycle of the parasite. The intermediate hosts are invaded by the eggs or early larval stages. Development of the larval stage occurs within the intermediate host. There may or may not be an asexual multiplicative phase within the intermediate host.

Larval stages escape from the intermediate hosts in one of two ways. Either they actively escape from the intermediate host and undergo a free-living stage before invading the next host; or they remain within the intermediate host, encyst and wait until the intermediate host is eaten by the next or definitive host.

Parasites that have evolved the requirement for more than one host will only survive if the two hosts share the same habitat and have some close association with each other. One of the most common associations is that of a predator–prey relationship. Another frequently encountered connection is when the intermediate host feeds upon the body fluids of the definitive host (as in the case of the arthropod vector). It can be said that in such situations the intermediate host is itself a temporary parasite of the definitive host. In those parasites whose life-cycles involve more than one intermediate host, normally no multiplicative phase occurs within the second or even third intermediate host.

1.1.3 MULTIPLICATIVE PHASES

The probability of a single, free-living stage reaching a new host in most situations is relatively low. In most instances, contact with a new host occurs by chance. In order to increase the odds of reaching a new host the majority of parasites are very fecund, ie they can produce large numbers of eggs. In addition many can undergo multiplication during their developmental stages. Parasites have evolved many different methods of increasing their numbers and some of them are listed below:

- Eggs are swallowed from which larvae hatch out. The juveniles develop/mature into adult male and female forms. The female after copulation produces fertile embryonated eggs which pass out of the host eg *Ascaris* spp, *Toxocara canis*.
- Larvae are swallowed by the host and mature into adult male and female forms. The female produces numerous live larvae which remain within the host eg *Trichinella spiralis*.

- Larvae actively penetrate the definitive host, mature into adults and then produce eggs which exit via the faeces eg *Ancylostoma* spp (hookworms), *Schistosoma mansoni*.
- Larvae enter into the circulation via an arthropod vector and mature into males and females. The females release live larvae which enter the blood circulation and are transferred to another host via blood-feeding arthropods (vector hosts) eg *Wuchereria bancrofti*, *Brugia pahangi*.
- Larvae are introduced by a vector but do not enter the circulation. They mature in a subcutaneous location where they either actively escape, eg *Dracunculus medinensis*, or are removed by a feeding arthropod eg *Onchocerca* spp.
- A cyst containing several spores is swallowed and the host's digestive enzymes break down the wall or shell of the cyst to release the spores. These then invade host cells and undergo asexual multiplicative phases. After one or more asexual cycles, selected individuals develop into male and female gametes. The male fertilises the female ovum to produce a zygote and the nucleus of the zygote undergoes several nuclear divisions to produce spores. The outer layers of the zygote produce a protective covering membrane or shell and the structure now becomes a cyst which passes out of the host eg *Eimeria* spp.
- A vegetative stage such as a trophozoite is inoculated via a blood-feeding arthropod. Host cells are invaded, followed by an asexual multiplicative phase and then the production of male and female gametes. Fertilisation occurs when the gametes are taken up by a feeding vector eg *Plasmodium* spp.
- A vegetative stage such as an epimastigote is introduced during vector feeding. Host cells are invaded followed by an asexual phase but no sexual multiplication eg *Trypanosoma* spp.
- Eggs are swallowed by an intermediate host and the larvae hatch out in the gut. The larvae migrate through the gut wall into the body tissues. They then undergo growth and development into advanced larval stages. They become quiescent within the intermediate host tissues and wait to be consumed by the definitive host eg *Taenia solium*.
- Eggs are swallowed and develop as above but undergo a process of asexual multiplication within the intermediate host eg *Echinococcus granulosus*.

1.2 THE NATURE OF PARASITISM

Endoparasites have become adapted to living within hosts.

Parasites and in particular endoparasites have adapted to living in specific parts of the host. The more specific the site the greater is the probability that the host–parasite relationship is stable and old in terms of evolution. Although the host still attempts to remove the parasite, the latter has evolved a whole strategy of avoidance mechanisms and, as long as the parasite can survive to reproduce, the effect of its presence upon the host is of no consequence.

Migration within the host, other than moving to a predetermined site, is usually an indication that either the parasite has found itself in the wrong host or this is a relatively new host–parasite relationship. Host reactions toward the parasite are a strong enough stimulus to force the parasite to move, trying to find a protected site. Often the chosen site is one where the host's immune response is comparatively weak or non-existent, such as the brain or eye.

Humans, in evolutionary terms, are the youngest of the mammals that inhabit the earth. Domestic animals have come into existence mainly due to man selectively breeding certain species and they too could be considered very young from an evolutionary point of view.

■ BOX 1.3 EFFECTS OF PARASITES UPON THE HOST

It is generally not in the interest of the parasite to destroy the host, and from the parasite's point of view it must try and avoid the host's attempt to destroy it. Parasites are mostly invertebrates and in terms of evolution are very much older and probably first adapted to being parasites of larger invertebrates, followed by the first aquatic vertebrates, the land vertebrates and finally the mammals. In general a state of equilibrium has evolved between the host's continuous attempts to destroy the parasite and the parasite's avoidance mechanisms.

The invertebrate groups evolved before the vertebrates and, once the vertebrates came into existence, invertebrates were able in many cases to take advantage of this new potential habitat first by invading it and then adapting to it. There has been a relatively long period for adaptation and those invertebrates destined to become parasitic have been able to invade then adapt to each newly evolved species. The parasite appears to have had the advantage and presumably before the final stages of adaptation the host suffered and to a certain extent so did the parasite. Man and domestic animals are still being invaded and are suffering as a consequence.

Man and domestic animals are both gregarious and the hosts' gregariousness favours the distribution of parasites. The closer the hosts live together the easier it becomes for the infective stage to find a new host. Humans live in close association with one another as well as with domestic animals and hence there are many parasites that are able to live in both man and numerous other mammals. Zoonosis is the term applied the situation where a parasite lives in both man and other animals.

There are very few vertebrates that are free of parasites; and those parasites affecting the lives of both man and domestic animals are often referred to as 'parasites of economic importance' eg *Fasciola hepatica*, *Haemonchus contortus*, *Taenia solium*, *Eimeria tenella*, *Plasmodium falciparum*, *Trypanosoma brucei* etc. The majority of parasites of economic importance occur in the following groups of invertebrates; the protozoa, the platyhelminths, the nematodes and the arthropods. Most modern studies of the biology of parasites concentrate on the first three groups; the parasitic arthropods tend be studied mainly by entomologists.

The diagram in Fig. 1.2 demonstrates, from a theoretical point of view, how host–parasite relationships have evolved. In long established hosts the parasites have reached an equilibrium with their hosts. If the equilibrium is disturbed in any way the following could occur:

- If the balance moves in the direction of the parasite/s feeding then the host suffers and the parasite/s have a pathological effect upon the host.
- On the other hand if the balance shifts in favour of the ability to inhibit the parasite, the latter will be destroyed.
- There is no advantage to the parasite if the host dies as a result of its presence.
- The equilibrium is based on the parasite obtaining sufficient nutrients to survive, but not to deplete the host in any way.
- If the host can contain the parasite, then it will only experience a minimal amount of discomfort due to the parasite's presence.

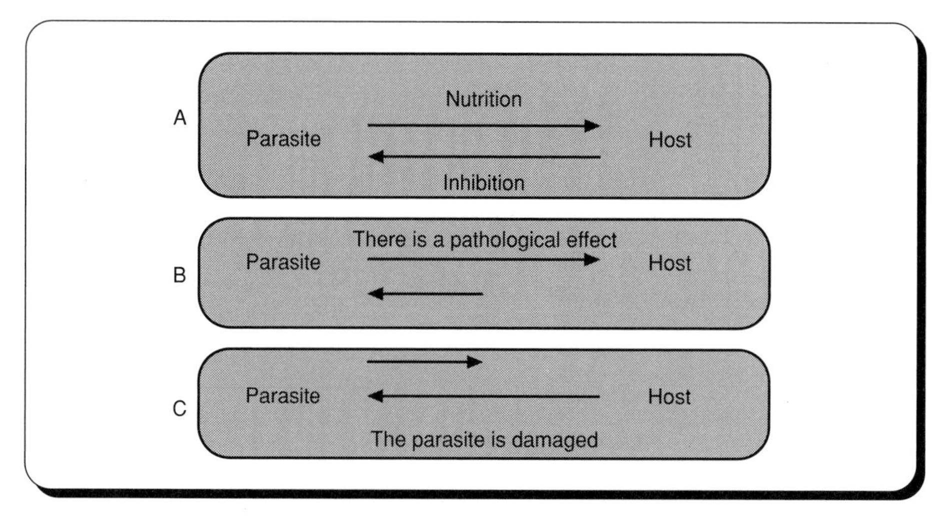

• **Figure 1.2** The survival of a parasite depends upon living in equilibrium with its host. The stability of the relationship is a balance between the feeding of the parasite and the host's ability to inhibit the parasite.
A: the equilibrium state.
B: pathological damage to the host resulting from the parasite's feeding.
C: damage to the parasite caused by the host's resistance.

The majority of established parasites have exploited their hosts to their maximum potential by making their hosts their environment. Parasites have adapted to almost every part of the vertebrate body, with some becoming so specialised that they will survive only in one particular niche within the host.

The majority of the parasites studied by parasitologists are endoparasites. They live entirely within the host, as opposed to ectoparasites which live on the surface of the host. Most ectoparasites are arthropods and tend to studied by people who specialise in animals from that phylum.

Ectoparasites live on the surface of their hosts.

The parasites of economic importance can effect the health and life of humans and domestic animals as well as the food supply. There are two main reasons why man and domestic animals tend to suffer from parasitic infection:

- Both man and domestic animals are gregarious, ie they live in relatively large communities close to one another. This favours the distribution of parasites.
- In evolutionary terms both man and domestic animals are very young.

1.3 PARASITES' EFFECT UPON THE HOST

The details of the pathological effects of selected parasites upon their host are covered in a later section. In this section only general considerations will be presented. In general the causes and types of damage done by parasites to their hosts are:

- Mechanical damage mainly due to blockages.
- Migration through tissues, penetration into cells.
- Nutrition; depleting the host of nutrients or competing with the host for essential nutrients.
- Toxins; very often the metabolic waste products of the parasite's metabolism accumulate in the host tissues and become toxic to the host.
- Immunosuppression; most parasites are relatively long-lived and continually present an antigenic challenge to the host, reducing the effectiveness of the immune response.
- Parasites in the 'wrong host' are in general more pathogenic. Most endoparasites have a preferred site within the host. This is not normally at the point of entry, hence the

need for migration. Parasites in the wrong host continue their migrations, 'lose their way' and end up in situations where they cause considerable necrosis of tissues.
- The physical presence of a parasite and its metabolic waste products can stimulate a negative host response, eg production of alkaline phosphatase, corticosteroids, etc.
- The age of the parasite; a young parasite may be able to avoid the host's response. An older parasite may die and the presence of the dead parasite may become problematic.

Any parasite once established within a host can become pathological in either a major or minor way. The site where the parasite finally settles will almost invariably be altered due to localised tissue damage caused by the parasite and the subsequent host reaction to control the parasite and repair the tissue damage.

Pathological damage to the host as a result of parasite invasion could be due to any one or a multiple combination of the following described below.

1.3.1 MECHANICAL DAMAGE

Once a parasite has actually penetrated a host, whether it is intracellular or extracellular (but within a tissue) or in one of the many body cavities (lumens), its physical presence will disturb the local homeostatic balance.

Physical features of the parasite such as hooks, protruding hardened structures and biting mouth parts damage the cells immediately surrounding the parasite.

Body fluids — including blood — and cellular debris accumulate around the parasite. The immediate effect is swelling and redness.

Apart from the gut and some of the lung-dwelling parasites, very few parasites are at their preferred site immediately after infection. The parasite then attempts to migrate from the site of infection to its predetermined site.

- Blockages; the filarial nematodes *Wuchereria bancrofti* and *Brugia pahangi* live in the subcutaneous tissue of the lower limbs. The result is oedema (a fluid buildup) of the lower limbs and also the enlargement of the genitalia.
- Heavy infections of *Ascaris lumbricoides* in children can block the small and large intestine.
- Erythrocytes infected with *Plasmodium falciparum* can become sequestered in the blood capillaries of the brain, causing cerebral malaria.
- The eggs of schistosomes can damage small blood vessels and block ducts within the portal system and blood vessels of the bladder.

1.3.2 MIGRATION WITHIN THE HOST

Parasites can move through the intercellular matrix between the cells and cell walls. The parasitic helminths progress through the tissues by active body movements, use of their mouth parts, hooks etc, and/or by the secretion of proteolytic enzymes. The larger the parasite the greater will be the physical damage to the tissue caused by the movement of the parasite.

The motile protozoan parasites use flagellae, cilia or an amoeboid type of locomotion to move through body fluids or tissue. Most protozoa generally travel via body fluids and often depend upon the blood circulation to reach their destination if it is not the blood itself. The intracellular parasites, once they have invaded their host cell, remain there until they reach a reproductive phase. To release the next phase into the host, the parasites burst out of the cell, which invariably destroys the host cell.

1.3.2.1 Possible reasons for parasite migration within the host

- The point of infection is not in the area of the parasite's organ or tissue.

The metacercariae of *Fasciola hepatica* are swallowed and once they excyst and the juvenile form is released into the gut lumen, it has first to penetrate through the gut wall to enter the body cavity. Once in the peritoneal cavity the young flukes move over the viscera until they reach the liver. They then burrow into a liver lobe and migrate through the hepatic tissue to enter the bile duct.

The tetrathyridia of *Mesocestoides corti* after been swallowed migrate through the intestinal wall. Some remain in the body cavity but others actively penetrate the liver tissue.

The infective larvae of the hookworms *Ancylostoma duodenale* undergo a process of *cutaneous larvae migrans* before they reach their definitive site in the gut.

- The developing stages or larval stages each require a different physiological environment to develop.

The eggs of *Ascaris lumbricoides* and *Toxocara canis* are swallowed and the larvae hatch out into the small intestine, the preferred site of the adult worms. Both types of larvae undergo a process of *visceral larvae migrans* and eventually return to the small intestine.

Plasmodium spp (those specific to humans) once inoculated into the bloodstream rapidly find their way into the liver cells. This type of migration is thought to be related to the parasite trying to find a safe hiding place from the host's immune defences.

The microfilaria larvae of filarian nematodes circulate round in the body fluids. They often can lodge in nervous tissue including the eyes.

1.3.3 THE PARASITE'S FEEDING ACTIVITY

Further aspects of parasite nutrition are dealt with in Chapter 6.

Feeding activities of the parasite can have a profound effect on the host by depleting or denying the host of essential nutrients and in the process causing damage to surrounding tissues.

Those parasitic helminths that have a gut and mouth feed upon the surrounding tissues, body fluids or even host cell debris which may have been caused by their presence. Parasites that have dietary preferences, such as mucosal blood, invariably evolve appropriate mouth parts to be able to puncture capillary blood vessels and can secrete anti-coagulants to maintain the blood flow.

The majority of the protozoa, the cestodes (the tapeworms) and most of the trematodes (the flukes) absorb their nutrients either through their cell membrane or via the outer tegument. They absorb molecules into their tissues from the surrounding host fluid and could deprive the host of certain essential nutrients.

In some instances, especially with gut parasites, they often compete with the host for essential nutrients. The tapeworm *Diphyllobothrium latum* can absorb vitamin B_{12} from the contents of the host's gut against a concentration gradient and deprive the host. *Schistosoma mansoni* feeds mainly on blood proteins; and a heavy infection can lead to the host suffering from malnutrition.

■ 1.3.4 TOXINS

Many of the toxins produced by parasites are waste products of their metabolism and are simply excreted and deposited into the host's fluids and tissues.

Haematin appears to be a metabolic waste product produced by *Plasmodium* feeding on haemoglobin. The haematin is deposited in liver and spleen where it apparently does have some harmful effects.

Amyloid deposits could be a reaction to certain parasite waste products.

■ 1.3.5 IMMUNOSUPPRESSION

Most parasitic infections are chronic and there is a continuous release of antigens into the host. The host's immune system has to either continuously react or become tolerant (see Chapter 5).

This condition can make the host more susceptible to normally lesser infections. Sheep with heavy infections of *Fasciola hepatica* often die because of the bacterium *Clostridium oedematiens* rather than as a direct result of the parasite.

Polyparasitism is a condition where the host is infected with more than one type of parasite eg different helminth species and/or a protozoan parasite.

Polyparasitism is quite common in humans, but only occurs once an established parasite has become a chronic infection. Measles epidemics are often associated with malaria endemic regions.

■ 1.4 HOST–PARASITE REACTIONS

The study of host reactions examines two types of hosts:

- A susceptible host is a host in which the parasite survives and the host may suffer as result of the presence of the parasite.
- A non-susceptible host is resistant to the parasite and either the parasite does not survive or exists only at a very low level of parasitaemia.

There is a range of different host responses and reactions to parasites. Some are simply mentioned below and are dealt with in detail in Chapter 5.

A host response may be altered after the experience of a primary infection. For example it has been observed that children who survive cerebral malaria develop some form of immunity. It is now thought that a low level *Plasmodium* parasitaemia provides a form of protection (non-sterile immunity). However this condition could also act as a reservoir for the parasites and maintain a population of infected vectors.

Schistosoma is also a parasite where living infection is thought to provide protection against challenge infections (molecular mimicry).

A healthy normal host reacts to primary invasion of parasite (non-self) material in a non-specific manner (non-specific or innate immunity). The parasite has evolved various ways of protection against this type of host reaction.

Plasmodium falciparum as rapidly as possible leaves the blood and invades hepatic cells to escape the 'phagocytes'.

Leishmania invades the macrophages and can avoid being digested.

The larger helminth parasites in general are able to escape phagocytosis. They usually become established before the host reaction can be effective. The majority of parasitic infections develop into chronic infections and instigate an adaptive immune response.

The details of how the response is initiated and develops are dealt with in Chapter 5.

■ 1.4.1 PATHOLOGY DUE TO HOST RESPONSE

There are only few diseases caused by the types of parasites already outlined that result in the death of the host. That is: not many parasites can be considered to be 'killers'.

It is of no real advantage for the parasite to kill the host.

All multicellular organisms (the host) react against internalised non-self material (the parasite). The host and the parasite have different genomes and hence each possesses a cell membrane with unique molecular structures. Recognition of this difference leads to specialised host cells attacking the invader. This first reaction leads to non-specific inflammation which is an innate immune response.

If the infection is localised near to the surface, skin inflammation results in localised redness, swelling and pain. However this also occurs when the infection is internalised and, although not visible, the pain will make the host aware of its presence.

If the immediate innate immune response does resolve the infection, this leads to the development of an adaptive immune response. One of the end products of an immune response is the formation of a granuloma. This basically consists of a cellular reaction to the presence of the parasite (see Chapter 5 for details). The parasite is eventually surrounded by fibrous cells and as it dies off, calcium is deposited within the cells and the physical presence of such a structure may block fluid flow and lead to necrosis of surrounding tissues.

■ 1.4.2 PATHOLOGY DUE TO PARASITE NUMBERS

Protozoan parasites undergo a multiplicative phase at some stage. This exaggerates all of the pathological effects already described.

The number of adult helminth parasites is generally dependent upon the number of infective stages that actually invade the host. The greater the number of infective stages that enter the host the greater will be the pathological damage.

The majority of helminths that reach sexual maturity within the definitive host reproduce, producing either numerous eggs or larvae. These later stages can also become pathological but have a different effect upon the host compared to that of the adult parasite. Hence it is possible that a single or pair of adult parasites releasing numerous reproductive stages such as eggs or motile larvae will have a considerable effect upon the health of the host.

There are relatively few, but significant, parasitic helminths that invade the host as eggs and once inside the host hatch into larval stages and then undergo a multiplicative phase eg *Echinoccus granulosus*. This results in the larval stages remaining for considerable lengths of time within the definitive host. A host reaction builds up around the parasites and the physical presence of such a structure can cause blockages and necrosis of surrounding tissues leading to a slow deterioration of the host.

■ 1.5 PARASITE LIFE-CYCLES

Two reproductive phases can occur during a parasite's life history:

- An asexual phase which results in the accumulation of numerous individuals of a particular species in one host. In some species, particularly among the trematodes, part of this process is considered to be parthenogenic rather than asexual.
- A sexual phase: the mature phase of the parasite's life history.

There are certain parasites which can undergo the processes of both sexual and asexual multiplication. Very seldom do both processes occur at the same time and they represent different phases of the life history. It is often referred to as 'alternation of generations' but unlike a similar phenomenon in plants there is no change in the chromosome number. Both phases or generations are normally diploid, the only haploid individuals being the gametes. To avoid confusion, all non-sexual reproduction will be referred to as asexual. If there is an increase in individuals within a single host, this will referred to as internal accumulation.

There are parasites that have a free-living phase outside the host. Some of these phases are active and actually reproduce, while others are quiescent.

The different types of parasite life-cycles can then be classified together according to the following criteria:

- The presence or absence of sexual reproduction.
- The presence or absence of asexual multiplication.
- Change and number of hosts.
- Internal accumulation of individuals due to either sexual or asexual reproduction.
- Whether or not there is an active or quiescent free-living phase.

1.5.1 ONE HOST ONLY, NO ASEXUAL MULTIPLICATION AND NO INTERNAL ACCUMULATION

When the host is infected, the parasite grows and develops. Once it has mature reproductive organs, sexual reproduction takes place. The resultant eggs, or if it is ovaviparous the larvae, escape into the external environment for dispersal and a 'chance' encounter with a new host.

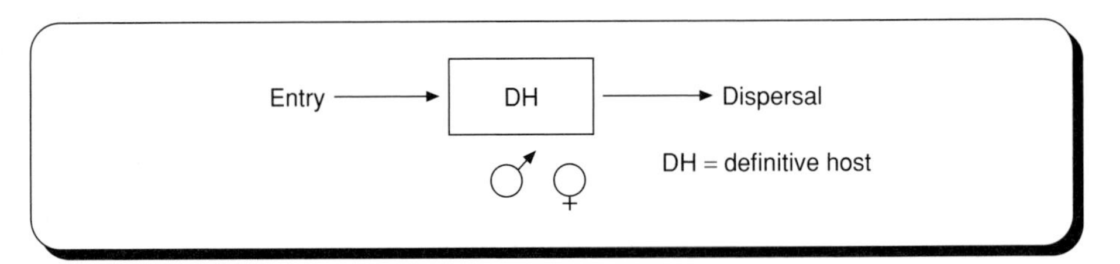

• **Figure 1.3** A simple direct life-cycle involves only one host. Within the definitive host (the host in which sexual reproduction occurs (DH)) are both male and female parasites. The eggs/larvae escape from the host to infect a new host.

All the Monogenoidea, trematodes and many of the nematodes eg *Ascaris* spp and the cestode *Hymenolepis nana* belong to this category.

H. nana is thought to have a simplification of a more complicated life-cycle. The stage normally associated with the intermediate host (the metacestode) develops within the definitive host. Eggs are swallowed, the oncosphere (the embryo) hatches out and penetrates the intestinal villi and grows into a diminutive cysticercoid (the metacestode stage). Within 4–5 days the cysticercoids emerge into the intestinal lumen and proceed to develop into adult tapeworms. The eggs pass out with the faeces.

Eugregarinida are parasitic protozoans in which sexual reproduction leads to the development of encysted zygotes, known as oocytes. Within each oocyte sporozoites develop. The function of the oocyte is to ensure dispersal.

1.5.2 LIFE-CYCLES IN WHICH NO SEXUAL REPRODUCTION HAS BEEN OBSERVED OR RECORDED BUT THERE IS ASEXUAL MULTIPLICATION AND INTERNAL ACCUMULATION

Entamoeba: the infective stage is a cyst which is accidentally swallowed. Within the gut lumen the cyst capsule is digested releasing 'juveniles' which invade cells and which then

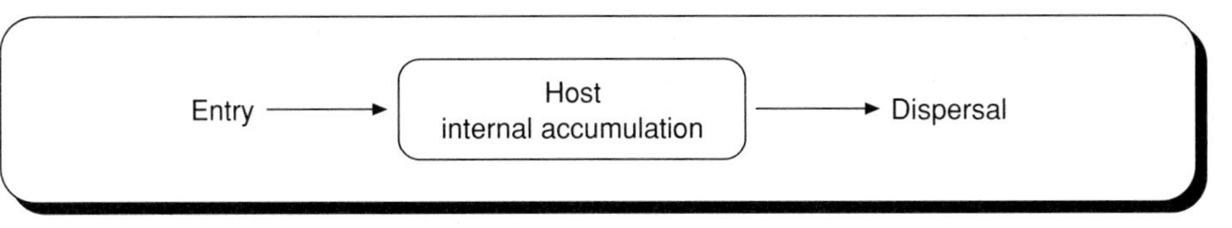

• **Figure 1.4** A direct life-cycle in which there is no apparent sexual reproduction. There is an accumulation of individuals within the host due to asexual reproduction before dispersal to the next host. (D = dispersal)

divide by fission. After a period of time some of the daughter parasites encyst and are then expelled from the host via the faeces for dispersal.

■ 1.5.3 LIFE-CYCLES WITH BOTH SEXUAL AND ASEXUAL REPRODUCTION WITHOUT CHANGE OF HOST AND WITH INTERNAL ACCUMULATION

Eimeria spp and *Isospora* spp (the Coccidia): the oocysts are swallowed and sporozoites are liberated into the gut lumen. The sporozoites then invade the cells of the intestinal epithelium and reproduce asexually by a process of schizogony which produces and then releases numerous merozoites. Released intercellular merozoites invade healthy epithelial cells and the process of schizogony is repeated. After at least one generation of asexual reproduction some of the merozoites transform into male and female gamonts. The nucleus of the male gamonts divide to produce numerous uninucleate biflagellate microgametes. The female gamont does not divide and remains as a uninucleate macrogamete. The male microgamete penetrates the female macrogamete and the two nuclei fuse to form a zygote. The zygote develops into an oocyst containing the next generation of infective sporozoites and is then expelled from the host.

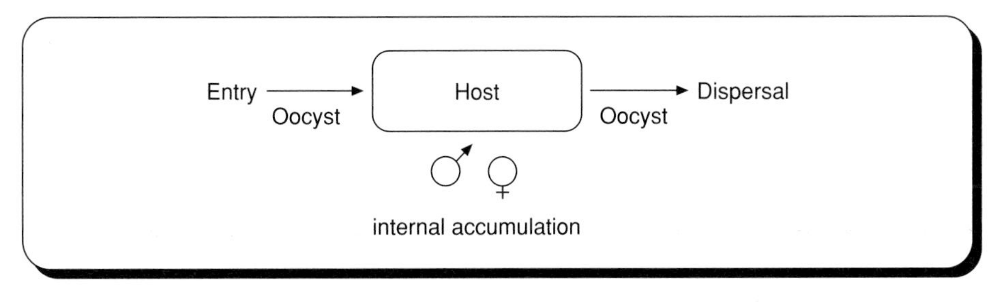

• **Figure 1.5** A direct life-cycle. The infective stage, the oocyst, invades the definitive host. There may be an asexual increase in somatic forms that gives rise to the male and female gamonts. Fusion between the male and female gametes produces an oocyst.

■ 1.5.4 LIFE-CYCLES WITH BOTH SEXUAL AND ASEXUAL REPRODUCTION WITHOUT CHANGE OF HOST BUT WITHOUT INTERNAL ACCUMULATION

The parasitic generation (endogenous) is represented by a parthenogenic female and the free-living generation (exogenous) consists of 'unisexual' individuals (see Fig. 1.6). The parasitic individual produces thin-shelled eggs which are either passed out via the faeces (*Strongyloides westeri*) or hatch in the gut mucosa into larvae (*S. stercoralis* or *Rhabditis* spp) which pass out of the host's gut. In the soil two types development occurs: homogonic (direct), or heterogonic (indirect):

- Homogonic. The larvae (rhabditoid type) develop directly into infective filariform larvae ready to reinvade a new host.
- Heterogonic. The larvae within 2–5 days develop into either mature male or female forms. After mating the females release rhabditoid eggs and the larvae hatching from them molt into infective filariform larvae.

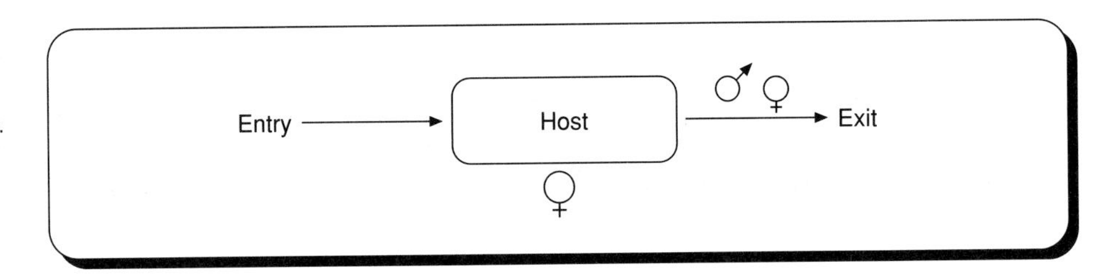

• **Figure 1.6** Only adults of one sex (usually parthenogenetic females) are found within the host. Both male and females develop during the free-living stage.

Female forms penetrate the intestinal villi and eventually produce eggs. Parasitic males are apparently unable to penetrate the mucosa.

1.5.5 PARASITES WITH A SEXUAL PHASE IN ONE HOST (THE DEFINITIVE HOST) AND LARVAL DEVELOPMENT BUT NO ASEXUAL OR INTERNAL ACCUMULATION IN THE INTERMEDIATE HOST

This type of life history (see Fig. 1.7) involves metamorphosis of the larva in a different host from that inhabited by the adult parasite:

- Cestodes (the tapeworms) — those that do not have an asexual multiplicative phase — for example *Taenia solium*, *T. saginata*, *Dipylidium caninum*.
- Nematodes such as *Dracunculus medinensis*; *Wuchereria bancrofti*.
- Almost of the Acanthocephala (spiny headed worms).

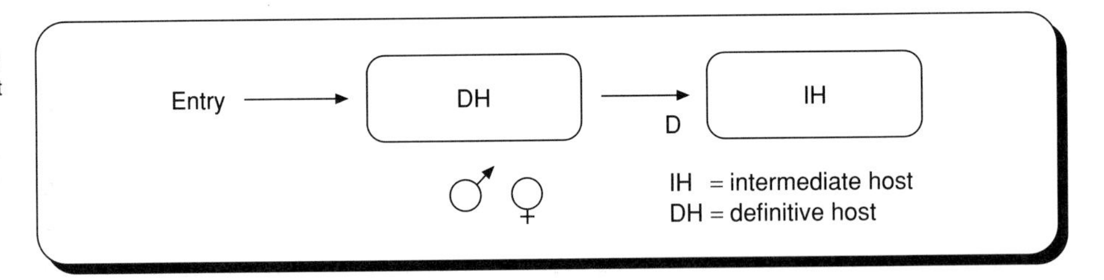

• **Figure 1.7** An indirect life-cycle involving at least two hosts, a definitive host and an intermediate host (IH). The eggs/larvae are released from the definitive host for dispersal. These stages infect the intermediate host from which further larval stages are released to invade the definitive host.

Adult cestodes are hermaphrodite and live within the gut of a vertebrate. Eggs are passed out via the faeces into the external environment — a dispersal phase. The eggs are eaten (accidentally) by the intermediate host. They hatch out and migrate into the host tissue and develop into the metacestodes stage. They remain within the intermediate host until it is eaten by the definitive host. These parasites have become adapted to a predator–prey relationship between the definitive host and the intermediate host.

The nematodes that have life-histories within this category have separate sexes but both are within the same definitive host. The fertilised females produce live larvae (microfilariae) which circulate round the body via the body fluids or migrate through the connective tissues. The larvae are dispersed to new hosts via vectors, usually a blood sucking arthropod.

Trichinella spiralis, a nematode — the mature adults live within the gut lumen. The female produces live larvae that migrate to striated muscles, the adults are expelled from the gut after about 14 days. The larvae encyst within the striated muscles and remain there until the host is eaten by the next host. In this example the same host is first the definitive host and then becomes the intermediate host.

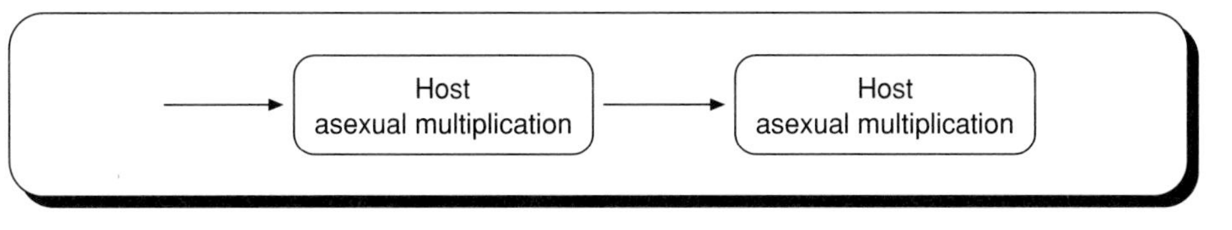

• **Figure 1.8** A two-host indirect life-cycle. Sexual reproduction has not been detected in either host. The parasites are capable of asexual reproduction in either or both hosts.

1.5.6 ASEXUAL REPRODUCTION WITH INTERNAL ACCUMULATION BUT NO SEXUAL REPRODUCTION AND NO EXTERNAL PHASE

- Trypanosomidae (extracellular blood parasites) are infected via arthropod vectors and reproduce asexually in both hosts and sometimes twice within one life-cycle.
- Piroplasmidae are intracellular blood parasites transmitted by *Ixodid* ticks. Asexual reproduction occurs within red blood cells of the vertebrate host and multiple fission takes place within the tick host, resulting in an accumulation of parasites within the tissues of the tick. The parasites are passed on to the next generation of ticks via the tick eggs.

1.5.7 LIFE-CYCLES WITH TWO CHANGES OF HOST BUT WITHOUT ANY INTERNAL ACCUMULATION

This occurs only in few acanthocephalans eg *Corynosoma stramosum*.

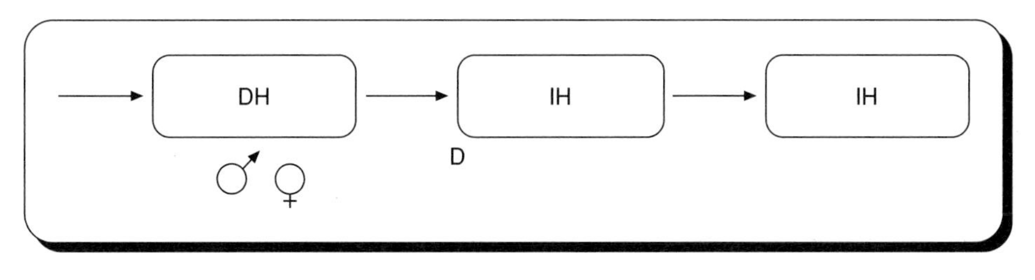

• **Figure 1.9** A three-host indirect life-cycle. One definitive host and two intermediate hosts (IH). There is a free-living dispersal stage between the definitive host and the first intermediate host. The second intermediate host is the prey of the definitive host.

1.5.8 LIFE-CYCLES THAT HAVE MORE THAN ONE HOST WITH BOTH SEXUAL AND ASEXUAL PHASES AND INTERNAL ACCUMULATION

This type of reproduction is found among a limited group of cestodes and the digenean trematodes:

- In the Cestoda, the metacestode stage undergoes asexual reproduction by budding or fission within the intermediate host eg *Taenia crassiceps*, *Mesocestoides corti*, *Echinoccocus granulosus*, *Multiceps multiceps*.
- A large number of metacestodes accumulate within the intermediate host. The main dispersal phase occurs with eggs passing out of the definitive host.
- Digenean trematodes eg *Fasciola hepatica*, *Schistosoma* spp. The adults live within the definitive host and produce eggs which pass out via the faeces. In fresh water motile larvae hatch from the eggs and penetrate a mollusc (intermediate host). They undergo a multiplicative phase and release cercaria from the mollusc which penetrate the definitive host.

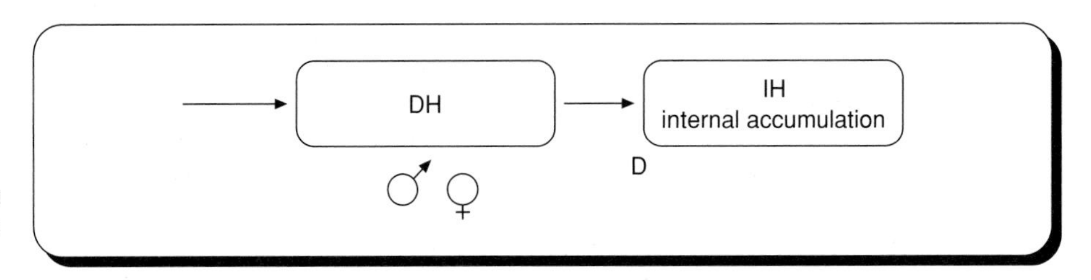

• **Figure 1.10** An egg/larva escapes from the definitive host into the environment for dispersal. Once inside the intermediate host the parasite undergoes asexual reproduction. The definitive host eats the intermediate host or the larvae escape from the intermediate host and remain free-living until they infect the definitive host.

1.5.9 LIFE-CYCLES USING THREE HOSTS

The trematode *Paragonimus westermani* adult lives in the lungs of dogs, cats and man. The eggs are either expelled from the throat or they pass via the trachea and oesophagus into the intestine where they pass out with the faeces.

From the eggs miracidia hatch out in fresh water. The miracidia penetrate snails of *Melania* spp. A multiplicative phase occurs within the 'liver' of the snail. First sporocysts are produced and from these emerges the next phase, the redia. The redia in turn give rise to cercaria which are released into the water and penetrate into a fresh water crustacean *Astacus japonicus* (a crayfish) or into a fresh water crab *Eriocheir japonicus*. In the muscle of the crustacean they encyst into metacerceria where they remain until the host is eaten.

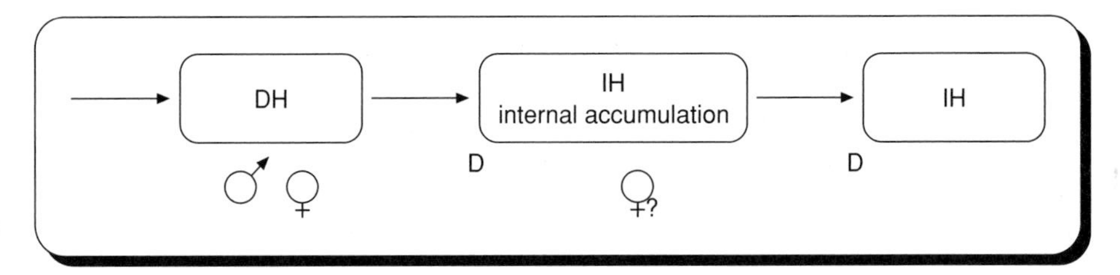

• **Figure 1.11** A three-host life-cycle with two free-living dispersal phases. Reproduction occurs within the first intermediate host, which may be interpreted as asexual or parthenogenetic. No reproduction occurs in the second intermediate host.

Ligula intestinalis is a pseudophylidean tapeworm, the adults live in the gut of fish-eating birds such as herons, pelicans, ducks, terns and gulls. Eggs pass out into water and hatch into a coracidium larva that is eaten by a crustacean, *Cyclops* spp, the first intermediate host. A procercoid develops within the *Cyclops* which is then eaten by a fish, *Rutilus rutilus*, the common roach, the second intermediate host in which the plerocercoid develops. The fish is then eaten by a bird, the definitive host, and develops into an adult worm.

Other similar examples are the cestodes *Schistocephalus solidus* and *Diphyllobothrium* spp, except that in the latter example fish-eating mammals tend to be the definitive hosts.

SUMMARY

The concepts of parasitism, what is a parasite and the different types of parasitism are discussed and described. The differences between a free-living animal and a parasite are emphasised. How parasites invade and escape from their host is outlined. The variations in parasite life-cycles are outlined using appropriate examples. There are parasites that have a direct life-cycle (ie only one host) and there are those that have more than one host (indirect life-cycles) and each parasite has a multiplicative phase which can occur at various stages during the life-cycle.

Humans and domestic animals are more likely to suffer ill effects from parasites because of the short time in evolutionary terms that parasite and host have been in contact. In order for both the parasite and the host to survive an equilibrium has to be established between acquisition of nutrients by the parasite and resistance to the presence of the parasite. There are various ways by which a parasite can cause pathological damage to the host and these are described with examples.

END OF CHAPTER QUESTIONS

INTRODUCTION TO PARASITOLOGY

Question 1.1 Why is parasitology considered to be a branch of ecology?

Question 1.2 What is the difference between a micro- and a macro-environment?

Question 1.3 What is the predetermined site?

Question 1.4 Which parasites are considered to be parasites of economic importance?

Question 1.5 In terms of parasitology what is the significance of a 'young' host–parasite relationship, in evolutionary terms?

Question 1.6 What is the main difference between a parasite and a free-living animal?

Question 1.7 Name and explain the different types of parasitism.

Question 1.8 Outline the methods parasites have developed for invading a new host.

Question 1.9 Outline the methods of escape from a host.

Question 1.10 Describe the different types of parasite life-cycles.

Question 1.11 What is the difference between a definitive and an intermediate host?

Question 1.12 Describe the various types of intermediate hosts.

Question 1.13 What is meant by a multiplicative phase?

Question 1.14 What are the different types of multiplicative phases observed among parasites?

Question 1.15 Discuss the significance of multiplicative phases from a parasite's point of view.

PARASITE–HOST INTERACTIONS

Question 1.1 What do you consider to be the reasons why man and domestic animals are more likely to suffer pathological damage as a result of a parasitic infection?

Question 1.2 What are zoonoses?

Question 1.3 What is meant by a host–parasite equilibrium?

Question 1.4 What happens when this equilibrium is disturbed?

Question 1.5 Outline the different effects a parasite can have upon its host.

Question 1.6 Using appropriate examples, describe what is meant by mechanical damage to parasites.

Question 1.7 Why do certain invasive stages have to undergo migrations within the host?

Question 1.8 What pathology is likely to result due to parasite migrations?

Question 1.9 What effects can parasite nutrition have upon the host?

Question 1.10 What are the toxins likely to be deposited as a result of the presence of a parasite?

Question 1.11 What is immunosuppression?

2 PROTOZOA

■ 2.1 THE CLASSIFICATION OF PARASITIC PROTOZOA

The protozoa are large collection of organisms that have only one thing in common in that they are all unicellular. They are all grouped together into a subkingdom which is divided into several phyla.

■ BOX 2.1 THE MAIN CHARACTERISTICS OF THE PROTOZOA

- Single celled or unicellular.
- Entire body bounded by a *plasmalemma*.
- The cytoplasm is bounded by a clear outer gelatinous region the *ectoplasm* and an inner more fluid region the *endoplasm*.
- Within the cytoplasm are the *organelles*; nucleus (or in some species nuclei), nucleolus, chromosomes, Golgi bodies, endoplasmic reticulum, lysosomes, centrioles, mitochondria and in some cases chloroplasts.
- Organelles unique to protozoa: contractile vacuoles, trichocysts and toxicysts.
- Contractile vacuoles are organelles involved in expelling water from cytoplasm, for volume regulation and osmotic regulation.
- Trichocysts develop within membrane-bound vesicles in the cytoplasm and end up in the periphery of the cytoplasm. Form into elongated capsules triggered by mechanical and/or chemical stimuli — discharge a long, thin filament.
- Toxicysts are related structures involved in predation — discharged filaments paralyse prey and initiate digestion.
  - Reproduction
- Asexual reproduction occurs in all groups and sexual reproduction in most groups.
- Asexual reproduction does not by definition generate new genotypes.
- Fission, a controlled mitotic replication of chromosomes and splitting of the parent into two or more parts.
- Binary fission, splitting of the individual into two.

- Multiple fission, numerous nuclear divisions precede rapid splitting of the cytoplasm into many individuals.
- Budding, a portion of the parent breaks off and differentiates into a new complete individual.
- Encystment, the individual loses its distinctive surface features such as cilia and flagella and rounds off. Excess water is pumped out by the contractile vacuole and a protective covering is secreted which hardens to form a protective cyst.
- Colonial, a single individual divides to form a colony of attached genetically identical individuals.
  - Feeding and digestion
- Digestion is entirely intracellular — ingestion is by phagocytosis or endocytosis.
- Food particles become surrounded by external membrane forming a food vacuole which moves about within the cytoplasm. The food vacuole is first acidic and then becomes basic. Digestive enzymes are secreted into the vacuole.
- Soluble nutrients are absorbed into the endoplasm. Solid wastes are discharged through an opening in the plasma membrane.

Most of the parasitic protozoa of economic importance (parasites of man and domestic animals) are found in the following phyla: the Sarcomastigophora and the Apicomplexa.

2.2 PARASITIC PROTOZOA OF ECONOMIC IMPORTANCE

Phylum Sarcomastigophora	
▪ Subphylum Mastigophora	
Class Zoomastigophora	
Order Kinetoplastida:	*Leishmania*; *Trypanosoma*
Order Dilpomomadida:	*Giardia*
Order Trichomodida:	*Trichomonas*; *Histomonas*
▪ Subphylum Sarcodina	
Superclass Rhizopoda	
Class Lobosea	
Order Amoebida:	*Entamoeba*; *Acanthamoeba*
Phylum Apicomplexa	
▪ Class Sporozoea	
Subclass Coccidia	
Order Eucoccidida	
Suborder Eimeriina:	*Eimeria*; *Isospora*; *Toxoplasma*; *Sarcoystis*
Suborder Haemosporina:	*Plasmodium*
Subclass Piroplasmia	
Order Piroplasmida:	*Babesia*; *Theileria*

2.3 FLAGELLATED PROTOZOA

Flagellates (mastigophorans) have an outer pellicle, definite body shape and one or more flagella (structure similar to cilia). Flagella often have mastigonemes (hair like projections) along their length.

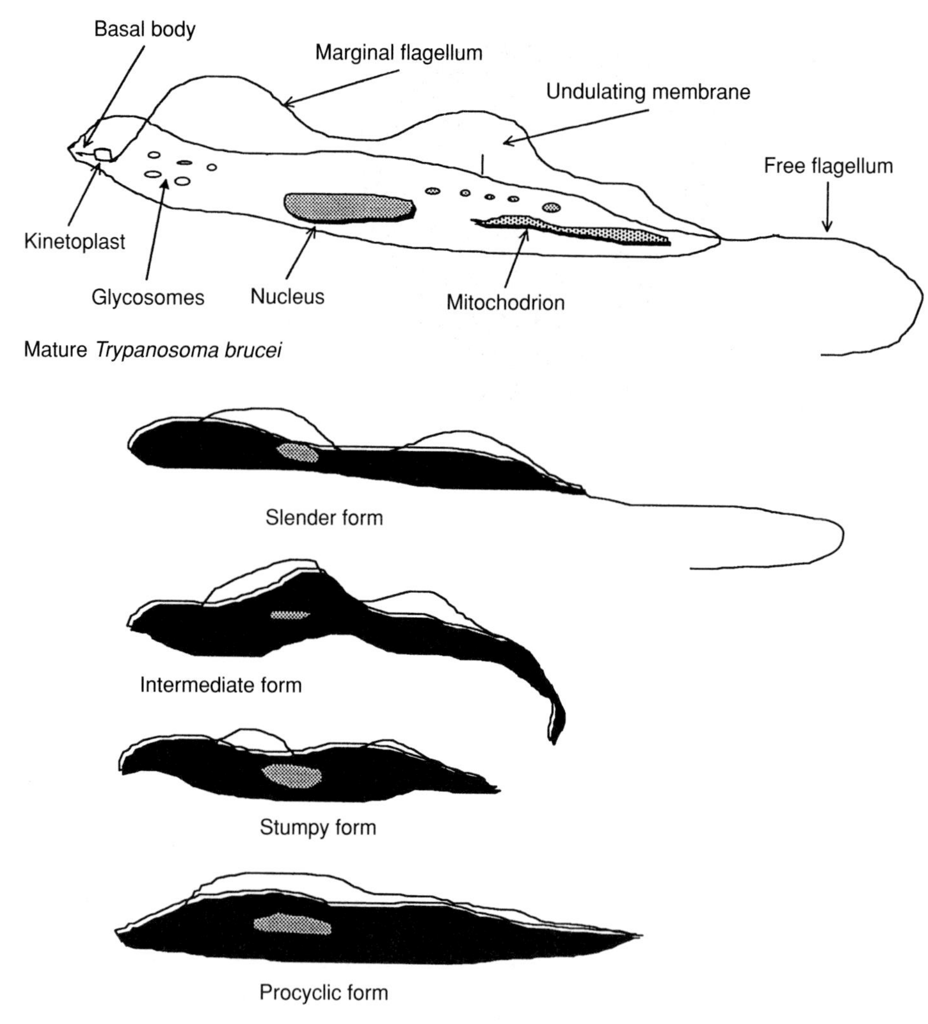

• **Figure 2.1** A mature *Trypanasoma brucei.* The African trypanosome has a single free flagellum, an undulating membrane, a single nucleus, a kinetoplast and basal body. The mature trypanomastigote form is normally found within the body fluids of the mammalian host. The other forms are part of the growth and development that occurs in both the mammalian and arthropod hosts.

Zooflagellated protozoa:

- Free-living and sessile forms in either marine or fresh water.
- Parasitic forms inhabit plants, invertebrates and vertebrates; often have complex life-cycles.

Approximately 25% of zooflagellates are parasitic. The majority have two hosts in their life-cycle. An intermediate host (normally a vector) and a definitive host. Examples of parastic zooflagellates:

- *Trypanosoma brucei*, causal agent of African sleeping sickness, see Fig. 2.1.
  Single flagellum and nucleus; large undulating membrane; a basal body (kinesotome) and a kinetoplast (a large mitochondrion-like structure); transmitted by the tsetse-fly; no sexual reproduction, only asexual reproduction.
- *Leishmania donovani*, transmitted by biting sandflies, causes skin damage to humans.

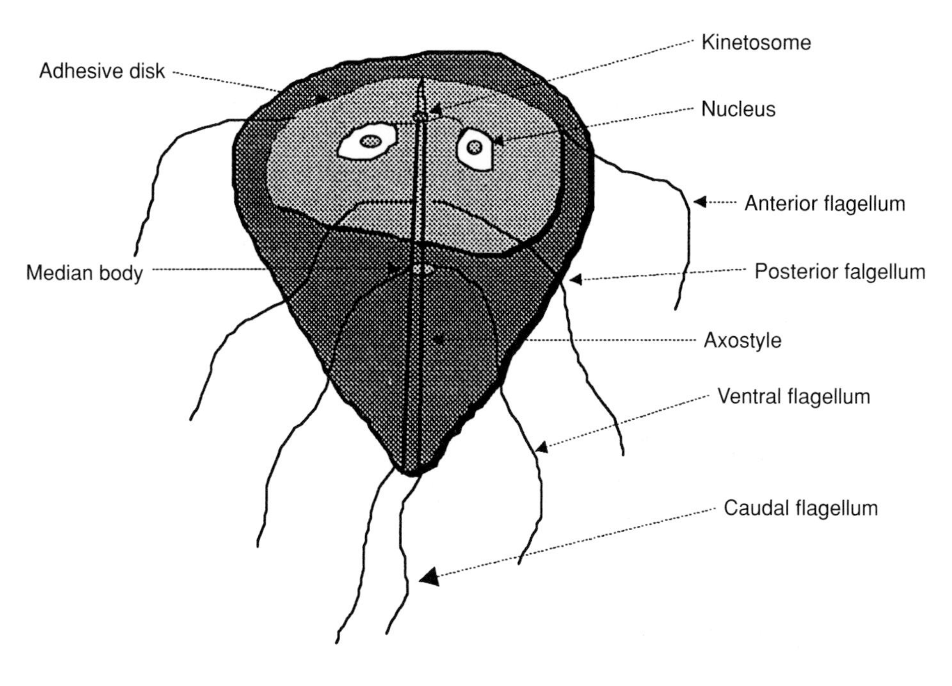

• **Figure 2.2** *Giardia intestinalis*, a multiflagellated binucleate protozoan with a central axostyle that may assist in maintaining a constant shape. The normal habitat of *Giardia* is the lumen of the intestine.

Multiflagellates

- *Trichomonas vaginalis* (single host), lives within the vagina, prostrate and urethra; transmitted via sexual intercourse.
- *Giardia lamblia* (single host), lives in gut; transmitted via drinking water, see Fig. 2.2.

The terminology used to name the various stages found in flagellates is based on the characteristics of the flagellum, its position and arrangement within the body, and the position of the kinetoplast and was proposed by Hoare and Wallace in 1966.

- Amastigote: represent rounded forms with no external flagellum, eg *Leishmania*.
- Promastigote: the flagellum arises from near to an antinuclear kinetoplast and emerges from the anterior end, eg *Leptommonas*.
- Opisthomastigote: these have a post-nuclear kinetoplast. The flagellum arises near it and emerges from the posterior end of the body, eg *Herpetomonas*.
- Epimastigote: the kinetoplast is alongside the nucleus, the flagellum arises near it, emerges from the side of the body and runs along a short undulating membrane, eg *Blastocrithidia* and stages of *Trypanosoma* and *Herpetomonas*.
- Trypanomastigote: these have a post-nuclear kinetoplast. The flagellum arises near it, emerges out from the side of the body and runs along an undulating membrane, eg *Trypanosoma*. The trypanomastigotes are the final developmental stage of the *Trypanosoma* and *Herpetomonas*.
- Choanomastigotes: these have an antinuclear kinetoplast. The flagellum arises from a funnel-shaped pocket and emerges from the anterior end of the body, eg *Crithidia*.

2.4 SPORE-FORMING PROTOZOANS

Phylum Apicomplexa, Class Coccidea eg *Eimeria*, *Plasmodium*, *Sarcocystis*, *Toxoplasma*, see Fig. 2.3.

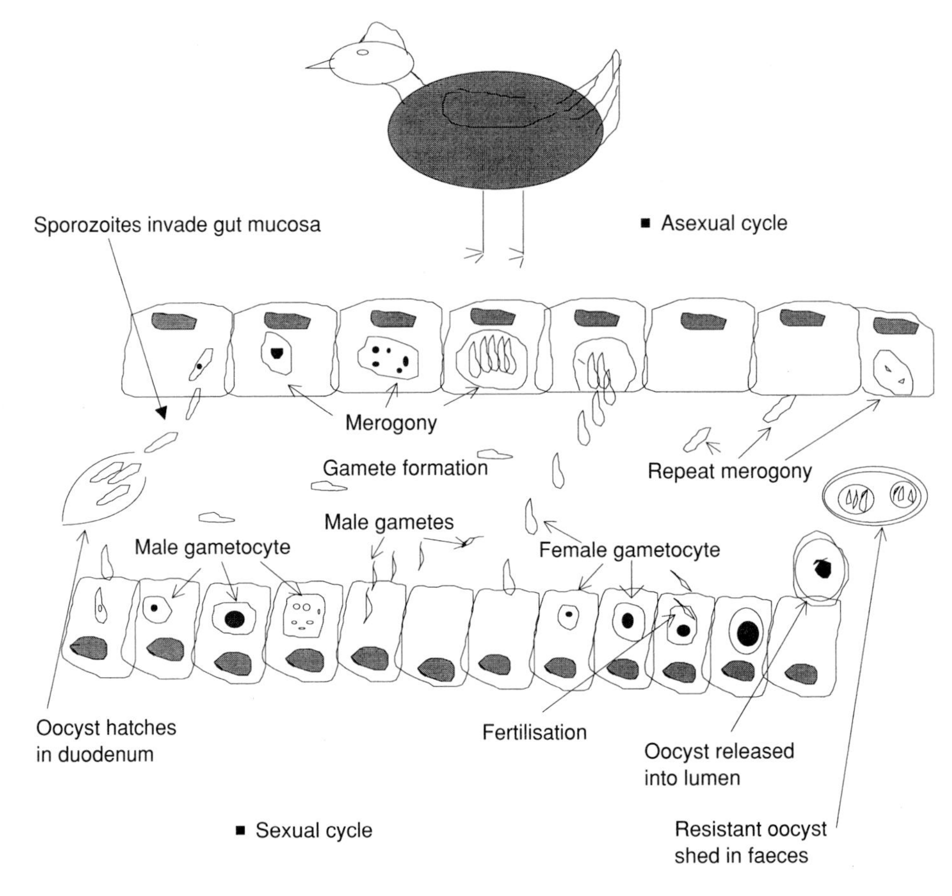

• **Figure 2.3** *Eimeria tennela*, a parasitic protozoan that is commonly found in the gut of domestic chickens causing the disease coccidiosis. The digested oocysts release sporozoites that then invade the epithelial cells of the gut wall and undergo asexual reproduction (merogony). After more than one asexual cycle male and female gametes are formed.

All the members of this phylum are endoparasites. At certain stages in the life-cycle all individuals possess an apical complex with a group of microtubules and organelles at one end of the cell and they all have flagellated gametes.

There are two main groups:

- The gregarines, parasites of invertebrates.
- The coccidia, parasites of both vertebrates and invertebrates, the latter usually serving as the vector or intermediate host. Many members of this group live in the red blood cells of vertebrates eg *Plasmodium* spp, see Fig. 2.4.

2.5 BIOLOGY OF PARASITIC PROTOZOA

The parasitic protozoa have like all endoparasites become adapted to a very specialised environment. However there are only a few morphological and anatomical adaptations that are unique to parasitic protozoa. The endoparasitic forms are either intracellular or extracellular.

The host cells that tend to be the most commonly parasitised are those rich in nutrients and with a high metabolic rate:

- Epithelial cells which absorb digested nutrients from the gut.
- Erythrocytes with their potentially high oxygen content and haemaglobin and hepatic cells rich in stored food.

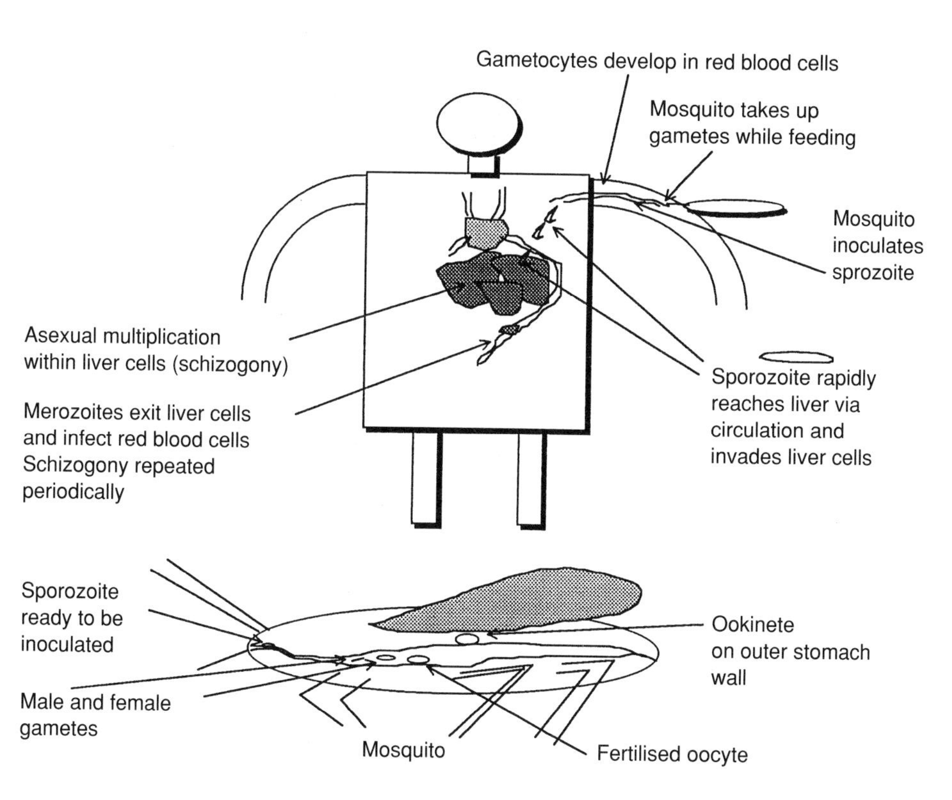

• **Figure 2.4** *Plasmodium* spp are parasitic blood-dwelling protozoa that cause the disease malaria. *P. falciparum*; *P. malariae*; *P. ovale* and *P. vivax* are the four species of *Plasmodium* that are transmitted to humans by female mosquitoes while feeding on blood. Sporozoites inoculated into the blood by a feeding mosquito are rapidly transported to the liver, invade hepatocytes and undergo asexual reproduction (schizogony).

In such cells the parasites undergo development, reproduction and escape to invade similar non-infected cells. Erythrocytes, and to a lesser extent epithelial cells, have a high turn-over rate. They are regularly replaced and hence it may be in the interest of the parasite's survival to produce new individuals to invade fresh cells.

Examples of parasites of the erythrocytes are the *Plasmodium* spp (the malaria parasites) and *Babesia* spp. The *Plasmodium* spp that infect humans and primates have an exoerythrocytic stage; immediately after the sporozoite is injected into the host it first invades hepatocytes (liver cells). In the liver cells there is rapid growth and development into a schizont which undergoes schizogony (a multiplicative stage, see Fig. 2.5). The liver cell ruptures and the newly formed merozoites are released and are temporally non-cellular but as soon as possible infect red blood cells (erythrocytes). The growth, development and multiplicative phase is repeated in the red blood cells.

Babesia spp — the vector host is a tick (*Boophilus* spp) and when an infective adult tick bites sporozoites are inoculated into the mammalian host. The sporozoites invade erythrocytes and undergo reproduction by binary fission and the newly formed merozoites invade non-infected red blood cells. There are no exoerythrocytic stages.

In both of the above examples the parasites have adapted to surviving in short-lived cells that are both numerous and continuously replaced. From the parasite's point of view this provides ample opportunity to increase its population and ensure that individuals survive to be transmitted to the next host.

Intestinal protozoan parasites such as *Eimeria tenella* enter their host via the oral route and once swallowed pass through the stomach into the small intestine, where they break out of their protective cyst. The first cells encountered by the infective stage of the parasite are those lining the lumen — the epithelial cells. Epithelial cells not only protect the

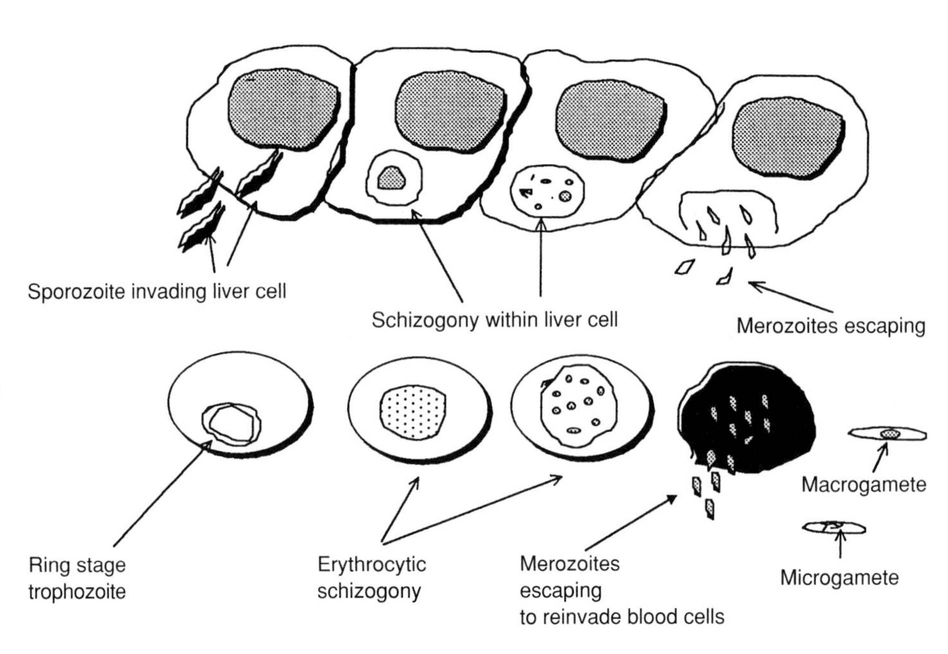

• **Figure 2.5** Merozoites released from the hepatocytes into the circulating blood invade erythrocytes (red blood cells) and undergo schizogony. This cycle is repeated more than once. Male and female gametes are formed during the blood phase and are taken up by a feeding mosquito. Fusion between the male and female gametes occurs within the mosquito gut.

gut but also function in absorbing digested soluble nutrients from the gut lumen. These cells are ideal for an intracellular gut parasite. The sporozoites, once inside the cell, develop and undergo reproduction by schizogony. The new generation bursts out of the cell and invades non-infected cells. This rapidly increases the number of individual parasites.

Parasites that enter the host, whether into the body fluids or the tissues, are vunerable to the host's defences and in particular the macrophages. These cells' prime function is to phagocytose (digest) non-host material and they are potentially lethal to any invading organisms. However, there are protozoans, eg *Leishmania* spp, that can invade and survive within macrophages. These parasites either resist or inhibit the production of destructive enzymes and inside a macrophage is probably the safest place for the parasites to live.

There are protozoa that are mostly extracellular in the mammalian host, eg *Trypanosoma brucei* (the African trypanosomes). These parasites have a single large flagellum and an undulating membrane. The parasites use both to propel themselves through the blood and they eventually exit from the blood to live in the cerebrospinal fluid. Extracellular parasites are continuously exposed to the host's immune responses and the African trypanosomes have evolved their own mechanism of avoidance. The host's attack is based on attempting to destroy the parasite's surface membrane but the parasite is able to regularly change the surface molecular configuration of its membrane in order not to be immediately recognisable to host immune mechanisms. This process is known as antigenic variation.

Trypanosoma cruzi the American trypanosome enters a mammalian host, via a vector, in the amastigote form and then pentetrates the reticulo-endothelial cell system or the muscles of the heart and commences an intracellular phase. In the muscle tissue the parasite forms a pseudocyst. After multiplication within the pseudocyst, the parasite transforms into a trypanomastigote form that escapes into the blood to begin an intercellular phase. Some of the circulating trypanosomes reinvade tissues and others are taken up by the vector when feeding.

T. cruzi has overcome two problems. Firstly it avoids the host's response by 'hiding' within cells where it multiplies and then invades their cells. Secondly it avoids the problem of distribution and transmission by invading the bloodstream.

Many protozoa take advantage of the fact that cells in tissue and organs are embedded in a semi-solid matrix — a ground substance. Most cells are adjacent to at least one other cell but they remain in contact with the fluids in the matrix to receive their nutrients, oxygen etc. and to dispose of their waste products. The majority of the infective stages of the parasitic protozoa are smaller than the host cells and are able move through matrix in search of an appropriate cell or enter the bloodstream through a fine capillary ending. The parasites move either by means of a flagellum or cilia or amoeboid-type movement.

The cellular architecture of most host organs is such that the individuals cells are in contact with intercellular fluid-filled spaces. Liver cells are arranged in columns two cells wide with an interstitial space between the columns. This provides ample room for the movement of phagocytic cells as well as parasites.

■ SUMMARY

An outline of the general characteristics of the protozoa is presented. The characteristics and taxonomy of the parasitic protozoa that infect mainly humans and domestic animals is described. Examples of the parasitic Sarcomastigophora, Sarcodina and the Apicomplexa are presented. A brief description is given of aspects of the morphology, physiology and reproduction of the most important examples of parasitic protozoa.

END OF CHAPTER QUESTIONS

Question 2.1 Give the names of the different types of outer coverings of protozoa.

Question 2.2 List the types of organelles found within the cytoplasm of protozoa.

Question 2.3 How do protozoa reproduce?

Question 2.4 How many different forms of asexual reproduction have been observed among the protozoa?

Question 2.5 What is a cyst; how and why do they form?

Question 2.6 How do protozoa feed?

Question 2.7 In which of the protozoan phyla are located the main parasites of economic importance?

Question 2.8 Name some examples of intercellular and extracellular parasitic protozoa.

Question 2.9 By what methods do parasitic protozoa move?

Question 2.10 What are spore-forming protozoa? Give some examples.

Question 2.11 Name some of the human diseases caused by parasitic protozoa.

Question 2.12 Name some flagellates that are the cause of diseases in both humans and domestic animals.

Question 2.13 Describe the process of schizogony.

Question 2.14 How many different forms occur during the life-history of an intracellular protozoan?

Question 2.15 Name and describe the different phases associated with the parasitic flagellated protozoa.

Question 2.16 What is the reticulo-endothelial system?

3 PLATYHELMINTHS

3.1 MAIN CHARACTERISTICS

- Three body layers: ectoderm, mesoderm and endoderm without a body cavity; aceolomate (without a body cavity)
  Main body structure:
- Acoelomate; dorso-ventrally flattened; bilaterally symmetrical; no anus.
- Mesoderm develops into a specialised connective layer known as the parenchyma.
- The organs — reproductive, nerves, muscular, excretory etc, are embedded within the parenchyma.
- No specialised respiratory organs. Gaseous exchange is by diffusion through outer body layer.
- Specialised excretory 'organs' known as flame cells unique to this phylum. These cells link into a tubular system for the removal of liquid nitrogenous waste.
- Nearly all members of this phylum are hermaphrodite.

3.2 TURBELLARIANS

Most of the turbellarians are free-living (about 3,000 species) and are found in marine, freshwater and moist terrestrial habitats. About 150 species are commensal or parasitic.

- They have ciliated outer layer.
- The average length is about 1 cm.
- Simple digestive system with variations within the group.
- They have a simple mouth opening with no well formed gut cavity. The food is packed into a dense mass of digestive cells.
- In the majority the mouth opens into a gut which may be straight, three-branched or multi-branched. More advanced species have a protrusible pharynx; some have a protrusible pharynx that can 'spear' food items.

3.3 CESTODES

A unique feature of this group is that the adults only occupy one type of habitat — the gut or derivatives such as the bile duct or pancreatic duct of a vertebrate.

- All adult forms are endoparasites inhabiting the gut or gut derivative of a vertebrate.
- They have no ciliated epidermis and no alimentary canal (gut).
- The adult form is divided into segments known as proglottids.
- The anterior end, the 'head', is known as the scolex and has well developed adhesive organs.
- With the exception of one family all are hermaphrodites.
- They all have a free-living stage and more than one host in their life-cycle.

3.4 TREMATODES (COMMONLY KNOWN AS FLUKES)

These are entirely ecto- or endo-parasites.

- No cilia on outer layer. The outer layer is known as tegument and is a living layer that helps to absorb nutrients.
- Body is undivided and has a mouth opening into a pharynx, oesophagus and then gut with a bilobed blind-ending diverticula (alimentary canal).
- Most species are hermaphrodites except for the Schistosomatidae.
- All have well developed adhesive organs.
- Two main orders: the Digenea and the Monogenea.

3.5 NEMATODES (COMMONLY KNOWN AS ROUND WORMS)

These organisms have three body layers with a body cavity known as the pseudo-coelome. The body cavity is fluid-filled and forms a hydrostatic skeleton.

- Cylindrical body shape, non-ciliated outer layer known as a cuticle.
- Sexes are separate; the gonads are tubular and with their ducts form a continuous structure.
- Worldwide distribution; occupying both terrestrial and aquatic habitats.

3.6 NEMATOMORPHA (COMMONLY KNOWN AS LARVIFORM WORMS)

These are non-segemented long thread-like free-living worms occurring in soil and water.

- Sexes are separate; adults are free-living but the larvae are parasitic in arthropods.

3.7 ACANTHOCEPHALA (COMMONLY KNOWN AS THE SPINY-HEADED WORMS)

These are unsegmented cylindrical worms, with a protrusible proboscis.

- No gut; pseudocoele; sexes are separate.
- Inhabit mainly the gut of vertebrates.

3.8 TAPEWORMS (EUCESTODA)

There are two main sub-classes of cestodes (see sections 3.8.2 and 3.8.3).

3.8.1 PROMINENT FEATURES

- The adult has an elongated tape-like body divided into proglottids.
- The absence of a gut in both the larval (metacestode) and adult stages.
- The outer covering is a living layer known as a tegument. All nutrients are absorbed through the metabolically active tegument, often referred to as naked cytoplasmic layer. It has the same structure in both the adult and metacestode stages.
- Apart from one group, *Dioecocestus* spp, all cestodes are hermaphrodites. They are monoecious ie they have both male and female organs and are protrandrous (males mature before the females).

3.8.2 CESTODARIA (A RELATIVELY PRIMITIVE SUB-CLASS)

The body is not divided into proglottids.

- Only one set of reproductive organs.
- No scolex; decacanth larva.
- eg *Amphilina foliacea*, whose definitive host is *Acipenser* (a sturgeon) and the intermediate host is *Gammarus* (a shrimp).

3.8.3 EUCESTODA (THE MAIN SUB-CLASS OF TAPEWORMS)

There are four common orders which are universally recognised: (1) Tetraphyllidea; (2) Trypanorhynca; (3) Pseudophyllidea; and (4) Cyclophyllidea.

The Eucestoda are divided into proglottids (except for one group, the Cariophyllidea). Each mature proglottid contains a complete set of reproductive organs. The eggs produced after fertilisation either embryonate in the uterus (known as an oncosphere larva) or in water (a coracidium larva). Both larvae have six hooks (hexacanth larvae).

3.8.3.1 The egg

The eggs are enclosed in a capsule or egg shell divided into an outer envelope, an inner envelope and an oncospheral membrane. There are two main groups of eggs: group 1 have an aquatic stage; and group 2 have no aquatic stage.

Group 1 eggs are laid in water and pass into an aquatic intermediate host. Most of the eucestodes except the Cyclophyllidea belong to this group. The Pseudophyllidea have an egg capsule with a 'lid' (operculate).

Group 2 eggs are embryonated and are laid normally onto soil and very rarely in water. The capsule when present is sometimes thin. The inner envelope is made of two zones: The outer zone (Zone I) is a cytoplasmic layer and the inner zone (Zone II) is a gelatinous layer. The innermost part of Zone II is the embrypohore.

3.8.3.2 The scolex

The anterior region, the scolex (the 'head' region), has attachment organs and is embedded into the gut mucosa (see Fig. 3.1). There are three types of attachment organs: bothria, bothridia and acetabula.

Bothria are narrow grooves with a limited amount of muscle; and they form sucking organs. Bothridia are mobile, stalked or sessile attachment organs and are mainly broad, flattened structures with thin flexible margins. They are mostly found in the Tetraphyllidea and Pseudophyllidea. Acetabula are sucking organs typically found in the

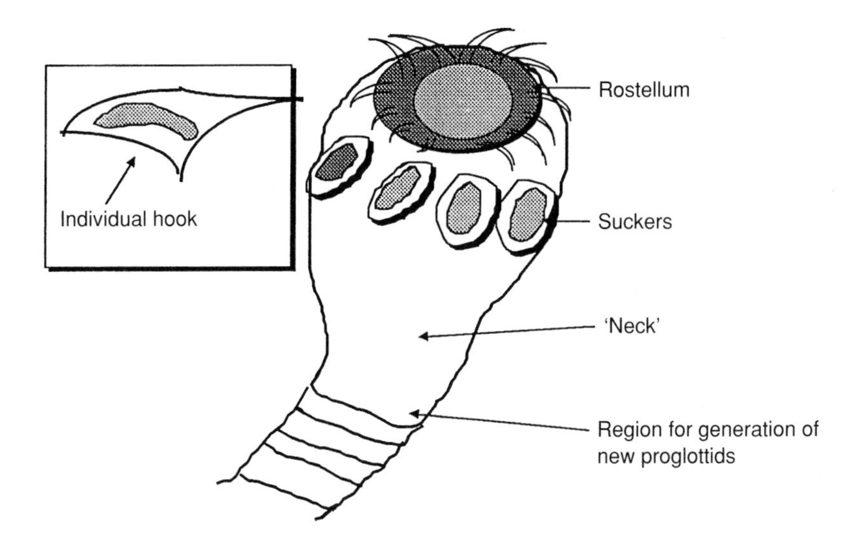

• **Figure 3.1** An adult tapeworm (cestode) normally inhabits the gut of a vertebrate. The scolex at the anterior ('head') end is equipped with an attachment 'organ'. The Cyclophyllidean cestodes have four suckers just below an apical rostellum. The rostellum has one or more rows of hooks. The shape and size of the hooks varies according to the individual species. Behind the scolex is the 'neck' region that generates the new proglottids.

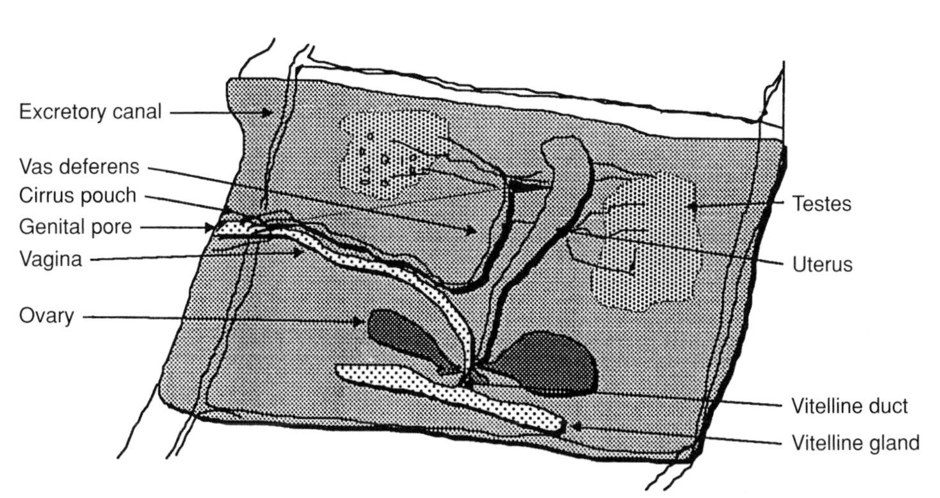

• **Figure 3.2** The main body of a tapeworm is made up of a series of repeating segments and a single segment is known as a proglottid. Each proglottid (except for the Dioecocestodidae) contains a complete set of both male and female reproductive organs embedded in a layer of parenchyma tissue.

Cyclophyllidea. At the very tip of the scolex of the Cyclophyllidea there may be a retractable ring of hooks, the rostellum. Behind the scolex is the neck region which is a regenerative region. This is where new proglottids are continuously formed to replace those lost at the posterior end. The chain of proglottids formed is known as a strobila.

3.8.3.3 The proglottid

The youngest proglottids are immediately behind the neck region. As they develop they appear to be separated from the one another by external constrictions.

Each proglottid (see Fig. 3.2) has an external living tegument, male and female reproductive organs, a lateral longitudinal nerve cord and a median longitudinal nerve cord (both are connected to a central nerve ring in the scolex). Scattered within the proglottid are flame cells linked to dorsal and ventral ('collecting') vessels which link up with vessels from neighbouring proglottids.

Mature proglottids become filled with eggs (gravid). There are two types of proglottid (sometimes called segments): (1) apolytic — the gravid proglottids are shed and pass

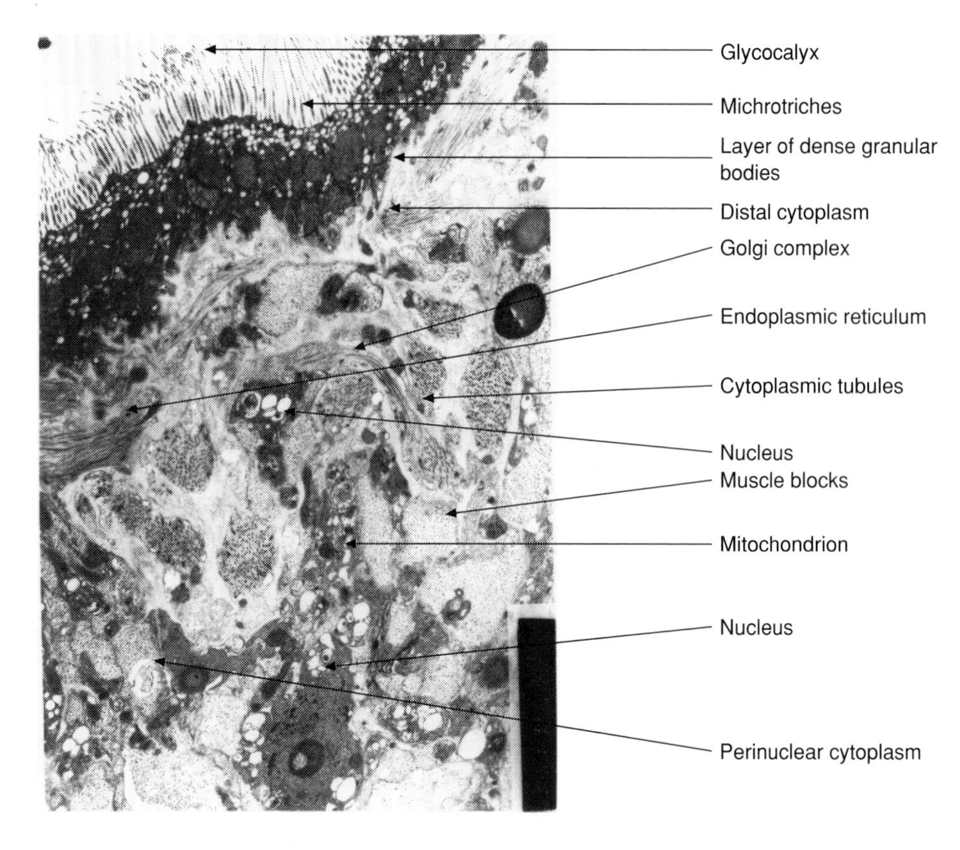

• **Figure 3.3** The outer layer of a cestode is a living layer known as a tegument. At the host–parasite interface is the glycocalyx. Microtriches extend the surface area. The tegument is a syncitium ie a non-cellular, multinucleate layer with the nuclei and organelles embedded in the inner layers of the tegument. The tegument is a metabolically active layer and all nutrients are absorbed through the tegument.

out with the faeces; and (2) anapolytic — the proglottids are detained and the eggs are released through a uterine pore.

3.8.3.4 The tegument

The tegument is the external covering of cestodes and trematodes and is made up of a syncitial protoplasmic layer divided into two zones: the distal cytoplasm and the perinuclear cytoplasm. The inner perinuclear cytoplasm is a nucleated layer containing tegumental cells and the distal cytoplasm is a syncitial layer which consists of a cytoplasmic extension of the perinuclear cytoplasm. Within the cytoplasmic layers are the organelles (mitochondria, Golgi bodies, endoplasmic reticulum), contractile vacuoles, dense secretory bodies and pinosomes.

The external surface of the distal cytoplasm has numerous micro-extensions known as the microthrix. This layer increases the surface area for absorption as well as possibly helping adhesion to gut wall.

A layer of mucopolysaccharides and glycoproteins forms a 'layer' — the glycocalyx — covering the microthrix. The glycocalyx can bind most enzymes and this may be the site of membrane or contact digestion and then the products of digestion are absorbed through the tegument (see Fig. 3.3).

3.8.3.5 The parenchyma

This is a network of mesenchymal cells that makes up the 'middle layer'. All the internal organs lie within this complex network of cells and the spaces within the network are filled with fluid containing glycogen, a food reserve.

3.8.3.6 The muscle structure

Beneath the tegument are two layers of myoblasts, muscle cells with lateral fibre-like outgrowths (fibrillae). One layer forms transverse or cuticular muscles with the fibrillae running transversely, the other layer forms longitudinal muscles (with longitudinal fibrillae). The muscle cells are made up of a non-contractile portion containing the nucleus and a myofibrillar portion that contains the myofilaments.

3.8.3.7 The nervous system

The major nerve ganglia are concentrated in the scolex. There are three nerve rings. The innermost, the central nerve ring, is the largest and links two lateral ganglia with a central ganglion. From the central nerve ganglion four median longitudinal nerve cords run posteriorly through the proglottids; and from the two lateral ganglia four lateral longitudinal nerve cords run backwards through the proglottids.

From each of the lateral ganglia arises an anterior longitudinal nerve. These join into the two rostellar ganglia which are linked by the rostellar nerve ring. The rostellar nerve ring supplies the rostellum with nerve cells and connects into the apical nerve ring (just beneath the rostellum).

The lateral ganglia also supply the suckers or adhesive organs with nerves and within the tegument there are numerous ciliated sensory receptors.

3.8.3.8 The excretory system

The basic functioning unit of the excretory system is a flame cell. These are cells in which one of the end cytoplasmic 'walls' is extruded into a flask-shaped long tubule that feeds into a collecting duct. The actual cell has a large nucleus with mitochondria in the cytoplasm. Cilia extend out of the cytoplasm into the lumen of the tubule; and the upper walls of the tubule carry a series of small projections, the leprotriches.

The beating of the cilia causes fluids that have seeped into the flask-shaped end of the cell into the tubular extension. The fluids then move away from the cell into the collecting ducts. This process removes excess liquid waste from the body, while a certain amount of reabsorption occurs along the walls of the collecting tubule. The liquid waste collects in a 'bladder' and then empties via an excretory pore.

3.8.3.9 The reproductive organs

Each proglottid contains a complete set of male and female reproductive organs. The only exception among the cestodes is found in the genus *Dioecocestus* where the male and female organs are found in separate individuals. The basic cestode arrangement of the organs in each proglottid is as follows:

Male The testes are paired and tend to be at either side of one end of the proglottid. From each testis emerges a collecting duct the *vas efferens* and these merge to form a single duct the *vas deferens*. The *vas deferens* leads into a short muscular tube, the cirrus, situated at one side of the proglottid in a pouch, the cirrus pouch, and this structure has an external opening, the genital pore.

Female There is either a single large ovary or a pair of ovaries linked into a common bulbous collecting tubule (see Fig. 3.4). There are several ducts exiting from the common tubule. Firstly there is the oviduct which ends in a coiled blind-ending uterus. On the opposite side is the duct which runs into the vagina situated in the cirrus pouch and opens to the exterior at the genital pore. At the bottom end of the common tubule is the vitelline duct that drains the vitelline glands. These supply material for egg membranes.

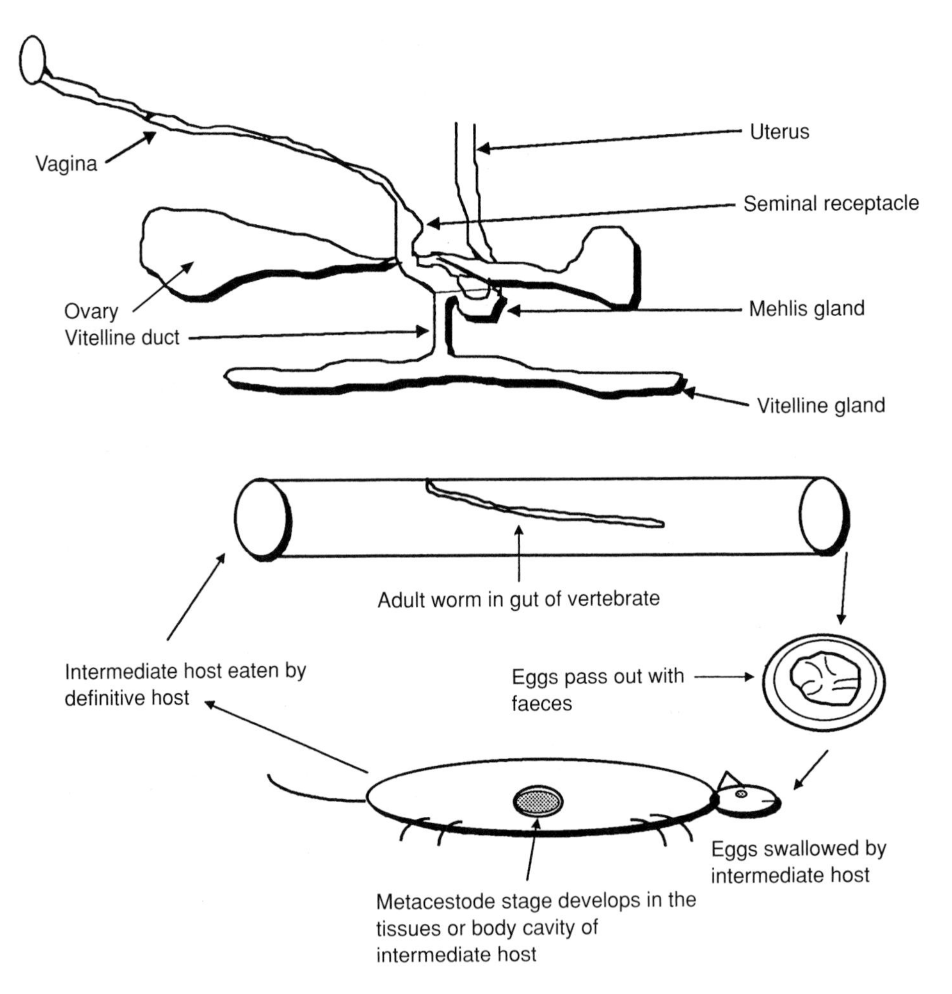

• **Figure 3.4** The details of the female reproductive system. The vitelline gland and the Mehlis gland are thought to be associated with the production of the egg membranes.

• **Figure 3.5** All adult tapeworms live in the gut or associated organs of vertebrates. The scolex becomes embedded within the gut mucosa and the rest of the 'body' lies unattached in the gut lumen. Cross-fertilisation between proglottids probably occurs. Eggs pass out via faeces and are eaten by the intermediate host. The larva hatches out of the egg and develops into a metacestode which remains within the intermediate host until it is eaten by the definitive host.

The cestodes are protandrous and it is thought that the cross-fertilisation takes place between proglottids.

■ 3.8.4 CESTODE LIFE-CYCLES

The cestodes require an intermediate host to complete their life-cycle. The eggs, having passed out into the environment via the faeces of the definitive host, are swallowed by the intermediate host. One major exception is *Hymenolepis nana* where, if the eggs are swallowed by the mammalian definitive host (mostly small rodents but man can also become infected), they can develop directly into adults and reinfect the same host or a similar host living in close association.

All types of vertebrates (homeothermic and poikilothermic) and arthropods are used by cestodes as intermediate hosts. There is far less host-specificity with regard to the intermediate hosts than there is with the definitive hosts. However there is a special and environmental relationship between the two types of host. The intermediate is almost invariably the prey of the definitive host (see Fig. 3.5).

The embryo that hatches out of the egg is conventionally referred to as the larva. The term metacestode was introduced in the early 1950s to describe the stage that the

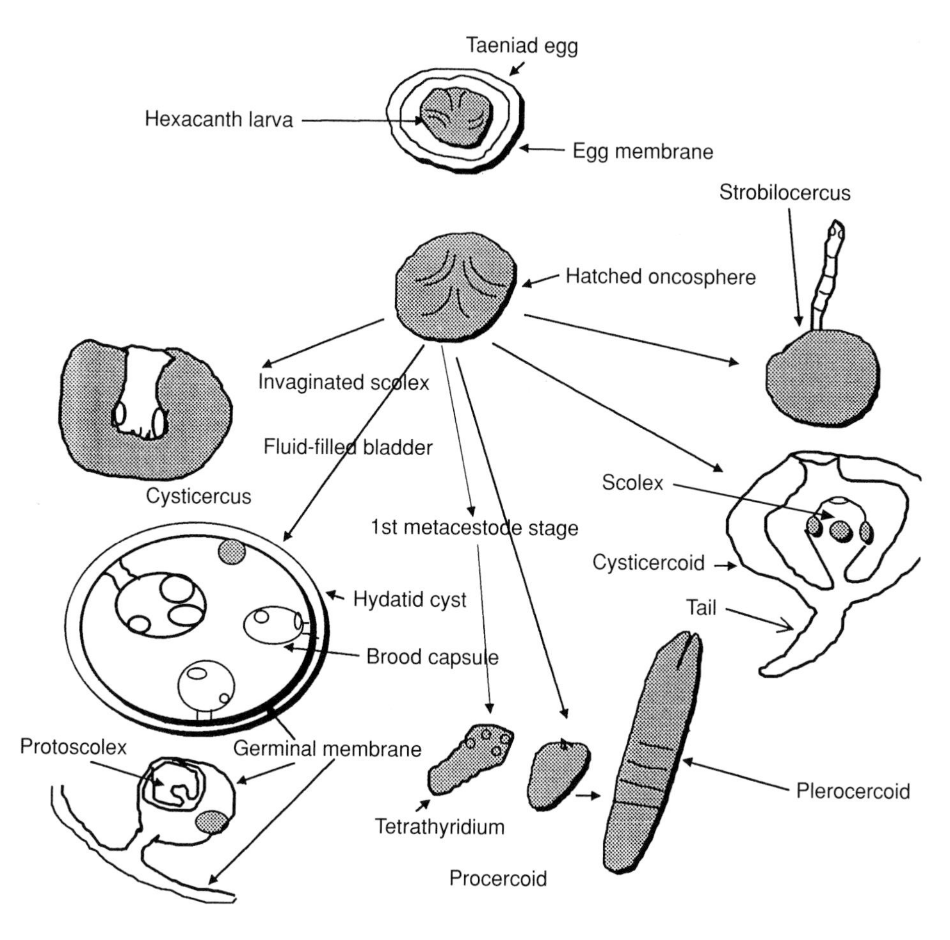

• **Figure 3.6** The larva hatches out of the egg and develops rapidly into a metacestode. Although throughout the cestodes there is very little variation in egg morphology, the development and morphology of the metacestodes differs in each individual cestode genus. Each type of metacestode occupies a characteristic niche within the intermediate host.

larvae develop into. (Metacestode is the term now most commonly used but there are some dissenters who claim that the use of the term larva is adequate and that to describe the developmental stages as metacestodes is unnecessary). Where more than one intermediate host is part of the life-cycle, the stage that lives in the second intermediate host is also called a metacestode.

Metacestodes occur in a number of different morphological types (see Fig. 3.6):

- *Procercoid* The eggs are swallowed mainly by crustacea and the larva hatches out and develops into a solid spindle-shaped structure with hooks on its posterior. This is typical of the pseudophyllideans and trypanorhynchids whose definitive hosts have some association with water.
- *Plerocercoid* This stage occurs in a second intermediate host and is a solid structure with scolex but does not have any hooks. Commonly found among the pseudophyllideans, trypanorhycans, tetraphyllideans and a few taenioids.
- *Cysticercoid* This a metacestode that nearly always develops in an insect, in particular one of the grain beetles (*Tenebrio* spp). It consists of a round anterior cyst-like vesicle within which is a non-invaginated scolex. There is a 'tail' arising from the posterior region which has hooks.

- *Cysticercus (Commonly known as a bladder worm)* This a fluid-filled cyst containing a single invaginated scolex and normally only develops in vertebrates, in particular herbivorous mammals. Some taenioid species produce external buds and hence undergo a form of asexual multiplication within the intermediate host.
- *Strobilocercus* This metacestode is found mainly in the liver of rodents. There is small bladder from which a solid structure extends with a scolex at its end.
- *Tetrathyridium* A small solid structure with four suckers and a scolex without hooks. Although normally a solid structure, it can under certain circumstances can become cystic and fluid-filled. In certain genera the tetrathyridia divide by longitudinal fission or by a process of budding.
- *Hydatid cyst* A fluid-filled cyst whose inner membrane gives rise to protoscoleces enclosed within vesicles known as brood capsules. The inner walls of the brood capsule continue to produce more protoscoleces. The formation of the brood capsules can be regarded as a form of internal budding.
- *Coenurus cyst* A fluid-filled cyst where the scoleces are formed in groups on the cyst wall. Secondary external cysts develop and usually remain attached to the original cyst to form a 'bunch of cysts'.

Whatever the form or shape of the larval or metacestode stage, they remain within the intermediate host until it is eaten by the future definitive host. When a metacestode is in its usual or most commonly encountered intermediate host, it nearly always migrates to a predetermined or preferred site. However when the metacestodes are swallowed by an animal other than the usual host, in many instances they still develop but may end up in 'unfamiliar sites' often causing considerable problems to this host. For example the cysticercus of *Taenia solium* is normally found in pig muscles, but in man it can migrate to the brain and cause neurocysticercosis. The hydatid cyst of *Echinoccocus granulosus* is the cause of hydatid disease in man.

3.9 CYCLOPHYLLIDEA OR TAENIOIDEA (TAPEWORMS)

These are parasites of birds and mammals. The scolex consists of four acetabula (suckers); there are no uterine pores; there is a single compact vitellarium posterior to the ovary.

Examples:

Taenia saginata (the beef tapeworm). Definitive host (the host in which sexual maturity occurs) is man and the intermediate host is cattle. Eggs pass from the gut onto grass and are eaten by grazing cattle. The larva hatches and develops into a metacestode, migrates through the gut wall and settles in striated and heart muscle. It develops into a cysticercus (bladderworm) and remains until eaten by man.

Taenia solium (the pork tapeworm). Definitive host is man and the intermediate host is pig. Cysticercus develops in striated and heart muscle.

Echinococcus granulosus. Definitive host is a carnivore (mainly domestic dogs) and the intermediate host can be man, sheep, horses, cattle, camels, etc. The adult is the smallest of the tapeworms — only three proglottids. Eggs swallowed by the intermediate host hatch in the small intestine. Metacestode migrates through gut wall into the abdominal cavity where it attaches to the liver or other viscera and develops into a hydatid cyst.

3.10 TREMATODES

There are three main orders: Aspidogastrea; Monogenea; Digenea.

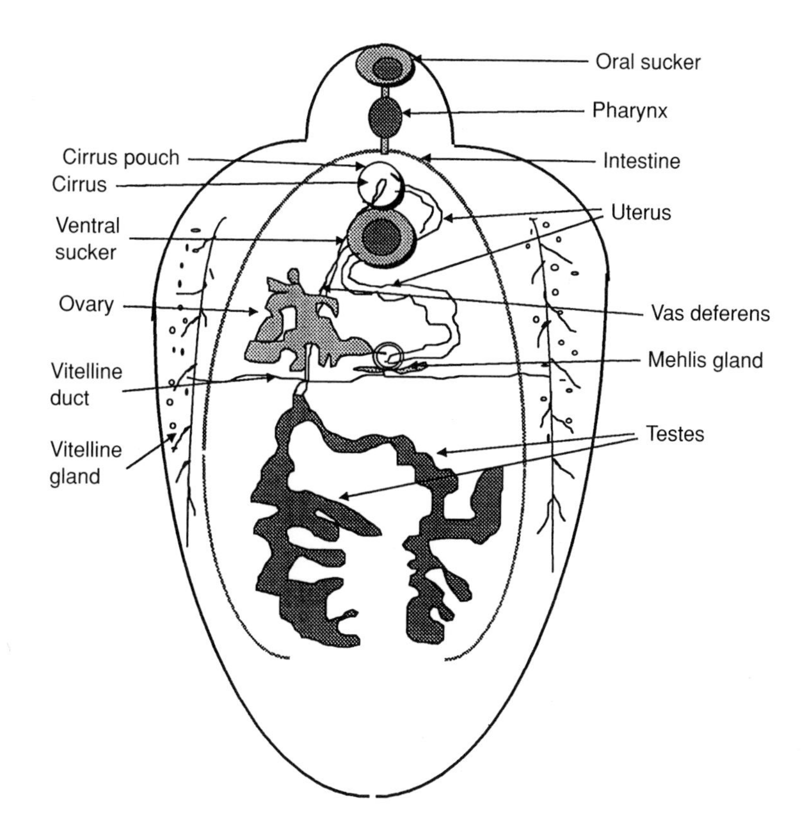

• **Figure 3.7** An adult trematode has a single flattened non-segmented structure and, apart from the Schistosomatidae, is hermaphrodite. They have an alimentary canal with a single opening leading into pharynx and gut diverticula. The reproductive organs are embedded in a layer of parenchyma tissue.

- Aspidogastrea are endoparasites with the entire ventral surface forming an adhesive organ.
- Monogenea are ecto-parasites of poikilotherms and have direct life-histories, ie involving only one host eg *Polystomma* spp. (The larval parasite is found in the gills of tadpoles and the adult migrates to the bladder of the frog.)
- Digenea are mainly endoparasites with one or two simple attachment organs and more than one host in the life-cycle. The intermediate host (apart from two exceptions) is always a mollusc eg *Fasciola hepatica*; *Schistosoma mansoni*; *Clonorchis sinensis*; *Paragonimus westermania*.

The adult digenean body consists of a single non-segmented oval or leaflike dorsoventrally flattened structure (see Fig. 3.7). The attachment organs are an oral sucker and a ventral sucker (also known as an acetabulum).

Adult digenea with rare exceptions occur exclusively within vertebrate hosts. The most common habitats are the gut, the lungs, the bile duct. One family, the Schistosomatidae, nearly all live in blood vessels. Other less common sites are the swim bladder, the eye and the urino-genital system.

■ 3.10.1 THE BODY STRUCTURE OF TREMATODES

All trematodes have a mouth, a muscular pharynx and a well developed alimentary canal, suitable for the digestion of semi-solid or viscous foods such as blood, bile fluid, mucous and intestinal contents.

The alimentary canal in digenea is a bifurcate structure. The mouth opens into the pharynx and this leads into an oesophagus and then into two blind-ending caeca. Each caecum may have many smaller branches extending out of it. The inner surface of the gut, the gastrodermis, both absorbs food and secretes digestive enzymes. All the organelles normally associated with active and passive absorption of digested food are present in the cells of the gastrodermis.

The outer covering of trematodes is a non-ciliated tegument which like that of the cestodes can absorb soluble nutrients and is also immunogenic. In structure and function the trematode tegument is very similar to that of the cestodes. The adult schistosomes live in blood vessels and an apparent adaptation to that environment is that the tegument (originally a trilaminate outer membrane) has become a heptalaminate membrane.

The excretory system is composed of numerous flame cells that empty into collecting tubules. The flame cells are also present in the cercaria and their number and arrangement vary with species.

The reproductive system is normally hermaphrodite, with two exceptions where the adults are unisexual: the Schistosomatidae and Didymozoidae.

- The male reproductive organs develop first (protandry). Two testes is the norm and two *vasa efferentia* lead from them into a single *vas deferens* ending in a cirrus opening into a genital sac. Surrounding the cirrus is a *ductus ejaculatoris* and a seminal vesicle. The pattern of spematogenesis is that 64 spermatozoa arise from a sperm bundle (a pattern common to most platyhelminths).
- The female has a single ovary and oviduct. Two vitelline ducts lead from the lateral vitelline glands into the oviduct which forms a chamber, the ootype (where the eggs are formed). Surrounding this area is a complex of glands known as Mehlis glands. A canal — Laurer's canal — leads from the oviduct and opens out on the dorsal surface. This structure is thought to be homologous with the vagina of the Cestoda.

The eggs are covered by a protective shell. In most of the digenea, monogenea and the pseudophyllidean cestodes the egg shell is made up of a 'tanned' protein sclerotin. The egg shell precursors have their origin in the vitelline glands (the vitellaria). Histochemical staining of the vitellaria shows that they contain the component proteins, phenols and polyphenol oxidase of the quinone-tanning system. Various functions have been described for Mehlis' gland and they include: lubrication for the passage of eggs; activation of spermatozoa; release of shell globules from the vitelline cells; activation of the quinone-tanning process; and creating a membrane to serve as a template on which shell droplets accumulate to form the egg-shell.

- The majority of eggs are non-embryonated and pass out into the environment via the faeces or urine, or in the case of lung infections via the sputum. Aquatic or moist conditions are required for the eggs to embryonate and develop.
- If the temperature, light, osmotic concentration and oxygen concentration are optimal, a motile miracidium hatches out into the water. The miracidium then swims around until it encounters the mollusc intermediate host. The eggs of *Dricocoelium dendriticum* hatch only when ingested by the intermediate host.

The miracidium The outer layer of the miracidium is covered with cilia. At the anterior end are three glands, two outer penetration glands and in the middle the apical gland. Below the glands there is a collection of nerve cells linked to two lateral nerves. Flame cells are scattered through the interior. There are two main types of cell: the somatic or body cells of the miracidium; and the germ cells which give rise to the next larval generation.

A miracidium hatches out from an egg, a stage common to most digenean life-cycles. In the majority of cases the miracidium emerges from the egg into water (except for *Dricocoelium* whose eggs are eaten by a snail) and with the aid of the cilia on its outer covering is able to actively swim for several hours. If it encounters the appropriate species of snail it penetrates into the snail's 'foot'; if not it dies.

Once inside the tissues of the snail, the miracidium migrates to the snail's digestive gland (the hepatopancreas) and there the miracidium's germ cells develop either into a sporocyst or directly into a redia. The germ cells within the sporocyst then give rise either to daughter sporocysts or directly to the next generation of redia.

Germ cells within the daughter sporocysts can develop in one of three ways:

- A free-swimming ceracaria which can penetrate the definitive host, eg *Schistosoma mansoni*.
- A metacercaria which encysts on within a second intermediate host, eg *Dricocoelium dendriticum*.
- A mesocaria which encysts into a metacercariae, eg *Alaria canis*.

The first generation of rediae are the mother rediae and these can develop in one of three ways:

- Develop into daughter rediae and their germ cells give rise to free-swimming cercariae and these then encyst on vegetation, eg *Fasciola hepatica*.
- Develop directly into free-swimming cercariae which penetrate the second intermediate host where they encyst, eg *Clonorchis sinensis*.
- Develop into ceracariae which are eaten by the definitive host, eg *Azygia longa*.

Rediae which arise directly from the miracidium become the mother redia and these develop along one of two ways:

- Develop into daughter rediae which produce free-swimming cercariae, which then become metacercariae, eg *Stichorcis subtriquetrus*.
- Develop into cercariae which transform into metacercaria, eg *Nanophyetus salmonicola*.

3.10.2 THE HABITAT OF ADULT TREMATODES

The habitat of the adult trematode (applicable to all endoparasites except for the type of intermediate host) must provide the following:

- Connections with the outside to enable eggs to pass out of the hosts' body. These can be any one of faeces, urine, blood or sputum.
- Provide a surface for attachment and also facilities for feeding.
- The environment must provide easily digestible nourishment to satisfy the high level of egg production.
- An intermediate host — a mollusc — which presumably provides some selective advantage related to the enormous asexual reproduction that occurs within a the mollusc.

Examples are *Fasciola hepatica* (the liver fluke; see section 7.5.7) and *Schistosoma* spp (see section 7.5.4).

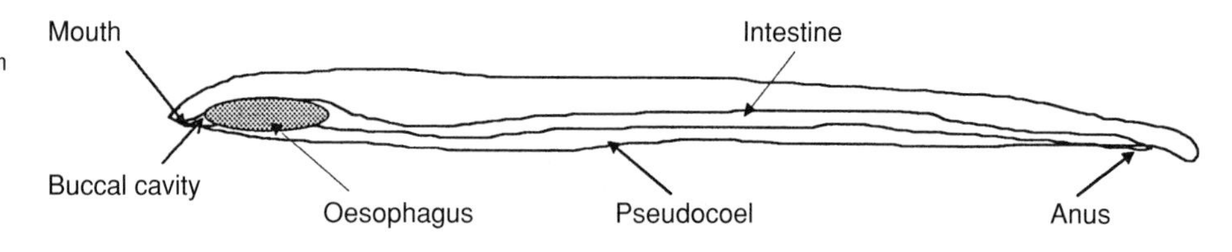

• **Figure 3.8** A nematode is a round cylindrical worm with a fluid-filled body cavity, the pseudocoel. At the anterior end is an opening, the mouth, that leads into the buccal cavity, the pharynx and then into the intestine which runs the length of the body. The gut ends in an opening, the anus, at the posterior end.

3.11 NEMATODES

General characteristics:

- All nematodes have a non-segmented cylindrical body. They are triploblastic with a fluid-filled body cavity known as a pseudocoelome (not a true ceolome).
- They have both free-living and parasitic forms. The free-living forms inhabit fresh-water, marine and terrestrial environments. Most free-living forms are relatively small (only just visible to the naked eye).
- Adult parasitic forms are found in both plants and animals. The adult parasitic forms range in size from microscopic to 10 cm plus.

3.11.1 MORPHOLOGY

A typical nematode has an elongated cylindrical shape, limbless without cilia and no respiratory organs (see Fig. 3.8). A typical nematode has the following body openings:

- A mouth at the anterior end. Slightly behind are small openings for the excretory organs — the amphids.
- The female opening — a genital pore — midway along the body length.
- Near the posterior end is the anus, the rear opening of the alimentary canal.
- In the male the genital organ opens almost at same place as the anus.

3.11.2 INTERNAL ANATOMY

The internal structure of a nematode consists of two 'tubes', the alimentary canal and the reproductive organs, contained within a fluid-filled cavity, the pseudoceolome (see Fig. 3.9). The outer cuticle is a non-permeable, multi-layered structure. On the inner side of the cuticle is the hypodermis consisting of a layer of longitudinal muscles. In addition there are unique types of muscles cells which 'suspend' the gut and reproductive organs.

The mouth is located at the tip of the anterior end and opens into a buccal cavity. This leads into a 'muscular' pharynx which in some species is subdivided into a series of 'bulbs'. The internal shape tends to be triradiate. The structure of the mouth parts and the pharynx are adapted to the feeding habits of the different species. The pharynx opens into a simple tube-like intestine with a subterminal opening, the anus.

The nervous system consists of a series of nerve commisures that are located in the anterior 'head' region and surround the pharynx. A series of large ganglia form a 'brain'. A dorsal and ventral nerve chord and two lateral nerve chords lead out from the main anterior ganglia. These nerve chords run through the hypodermis of the cuticle.

3.11.3 REPRODUCTIVE SYSTEM

In most cases the sexes are separate. In those species that are hermaphrodite the male gonads mature first (protandry). There are also nematode species that are parthenogenic.

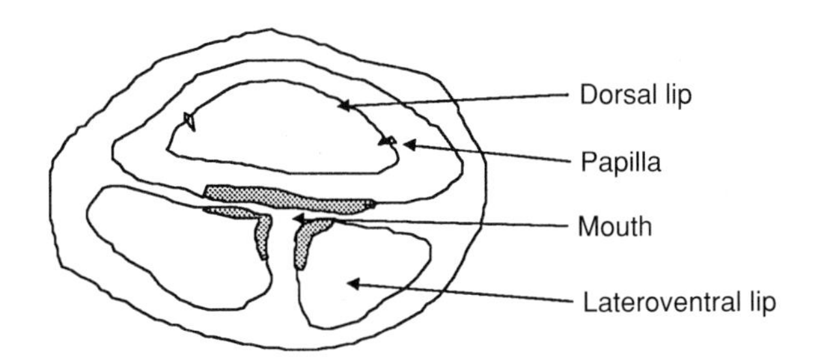

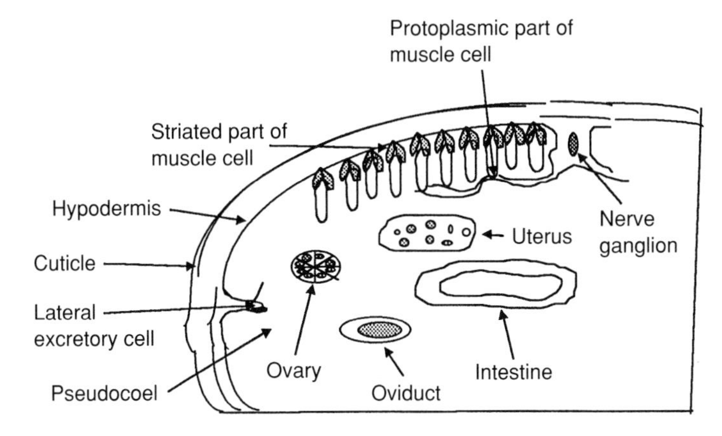

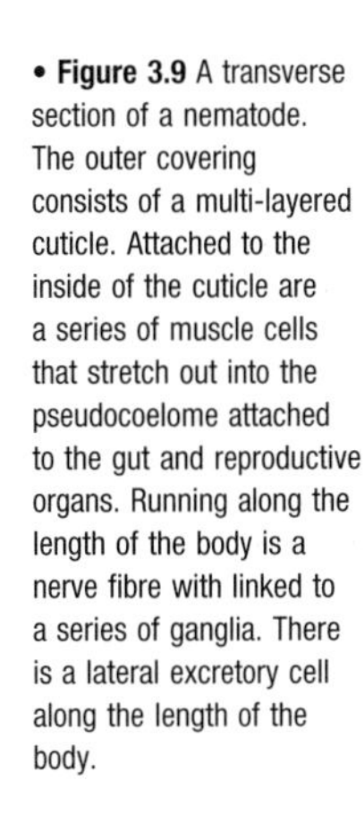

• **Figure 3.9** A transverse section of a nematode. The outer covering consists of a multi-layered cuticle. Attached to the inside of the cuticle are a series of muscle cells that stretch out into the pseudocoelome attached to the gut and reproductive organs. Running along the length of the body is a nerve fibre with linked to a series of ganglia. There is a lateral excretory cell along the length of the body.

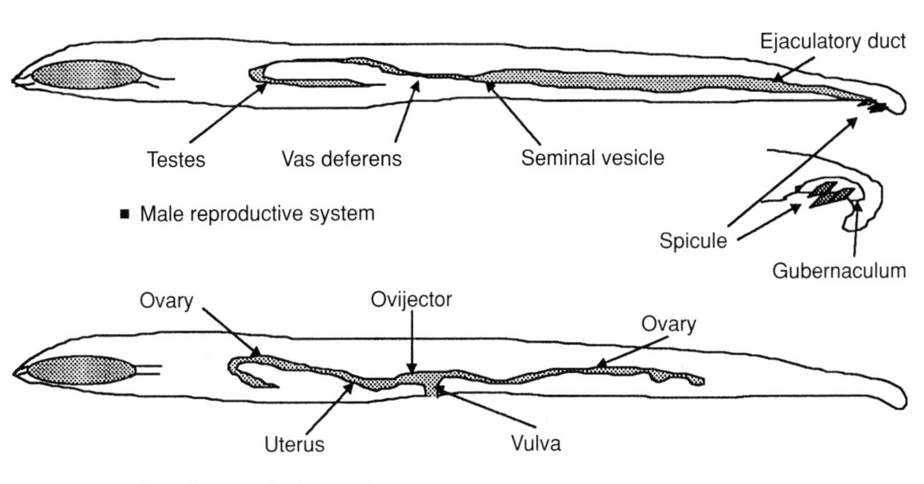

• **Figure 3.10** In nematodes the sexes are separate. The male organs consist of tubular testes linked, via a seminal vesicle, to an ejaculatory duct that opens into a cloaca. Along the cloacal opening are a series of spicules. The female organs consist of a pair of ovaries that open into the uterus, which has a lateral external opening, the vulva.

Males are smaller and possess copulatory aids. The male gonads consist of testes leading into a tube, the *vas efferens* and then into the *vas deferens* ending in the male opening.

Female gonads consist of ovaries that open into an oviduct leading into the uterus where the eggs are stored. The uterus opens into a vagina situated midway along the length of the female body (see Fig. 3.10).

■ 3.11.4 LIFE-CYCLE

The eggs hatch into a first stage larva. This moults into a second stage larva. There are usually four moults and the fifth stage usually develops into the adult.

■ BOX 3.1 EXAMPLES OF NEMATODES

Enterobius vermicularis Only one host during its life-cycle. The adult worms live in the gut. The female migrates to the perianal region at night to lay eggs. The eggs are swallowed (oral infection).

Ascaris lumbricoides Only one host during its life-cyle. The adult lives in the gut and the cycle begins with eggs being swallowed. Larvae hatch out of the eggs in the small intestine and migrate through the gut wall to enter the blood circulation. They migrate to the heart and lungs then back to the alimentary canal. During migration the larva moults and matures into the adult form within the gut.

Trichinella spiralis Adult male and female live in the gut. The female produces live larvae which then burrow into the intestinal wall and migrate to the striated muscle where they encyst. Both male and female adults are then expelled from the gut after 14 days. The larvae remain until the host is eaten and then they excyst and mature within the new host.

Wuchereria bancrofti The adult worms occur in tightly coiled masses in the lymphatic ducts. Female produces live larvae which enter into the blood circulation and are then transmitted to a new host by a mosquito.

Toxocara canis Pregnant infected bitch reactivates dormant L_3 larvae by hormonal changes during pregnancy. In a lactating bitch the larvae are passed to puppies via the milk. In bitch and puppies the larvae migrate from gut to liver, lungs and trachea. Coughed up and swallowed, they develop into L_4 larvae and adults and then release eggs.

■ SUMMARY

The characteristics and an outline classification of the Phylum Platyhelminthes are presented. The basic morphology and physiology of a typical cestode, trematode and nematode are described. The general plan of the life-cycles of each group is outlined. The cestodes have several different types of metacestode (larval stages) which occur within the intermediate hosts; and it is some of these that are the cause of severe pathology in humans and domestic animals.

The trematodes have an even more complex life-cycle but the intermediate host is always a species of mollusc. The basic life-cycle and variations that occur are outlined. The nematodes have a more regular pattern to their life-cycles. Normally the larvae moult four times into the next stage before reaching sexual maturity.

END OF CHAPTER QUESTIONS

PLATYHELMINTHS

Question 3.1 What are the distinguishing features of the phylum Platyhelminthes?

Question 3.2 How many body layers are there in platyhelminths and what are the layers called?

Question 3.3 What is the difference between a coelomate and an acoelomate animal?
Question 3.4 Name the different organs normally found in platyhelminths.
Question 3.5 How do these organisms 'respire'?
Question 3.6 How do platyhelminths remove liquid metabolic waste products?
Question 3.7 Name the classes within this phylum. What are the distinguishing features of each class?
Question 3.8 What is the difference between a nematode and a platyhelminth?
Question 3.9 How do nematodes maintain their shape and structure?
Question 3.10 What is an acanthocephalan and in what hosts are they most frequently found?

CESTODES

Question 3.1 What are the main features of cestodes?
Question 3.2 What are the main subclasses within the cestodes?
Question 3.3 Why are the Cestodaria considered to be a primitive group?
Question 3.4 Name with examples the four most common groups within the Eucestoda.
Question 3.5 Describe the characteristics of cestode eggs.
Question 3.6 What is the structure and function of a scolex?
Question 3.7 Name and describe the different types of attachment organs associated with a scolex.
Question 3.8 What is a proglottid and where is it formed?
Question 3.9 What organs are found in a mature proglottid?
Question 3.10 Describe the structure and function of the tegument.
Question 3.11 What types of tissues are found within the Cestoda?
Question 3.12 Describe the male and female reproductive organs.
Question 3.13 Outline a typical cestode life-cycle.
Question 3.14 Name and describe the different types of metacestodes.

TREMATODES

Question 3.1 Name with examples the three different orders within the class Trematoda.
Question 3.2 What are the differences between a monogenean and digenean trematode?
Question 3.3 What types of attachment organs are found associated with the trematodes?
Question 3.4 Describe the structure and variations of the digenean alimentary canal.
Question 3.5 How does the tegument of a digenean compare to that of a eucestode?
Question 3.6 Describe the digenean reproductive systems.
Question 3.7 What are the characteristic features of digenean eggs?
Question 3.8 Outline the basic life-cycle of both a monogean and a digenean trematode.
Question 3.9 Describe the larval stages that can occur during a digenean life-cycle.
Question 3.10 What are the necessary requirements of a trematode habitat?

NEMATODES

Question 3.1 What are the general characteristics of nematodes?
Question 3.2 Give a general description of a typical nematode.
Question 3.3 What are the distinguishing features of the internal anatomy of a nematode?
Question 3.4 How does a nematode maintain its body shape and what type of 'skeleton' does it have?
Question 3.5 Describe the mouth parts of a nematode in relation to its type of feeding.

Question 3.6	What types of musculature do nematodes have?
Question 3.7	Describe the nervous system and sense organs associated with nematodes.
Question 3.8	Outline the nematode reproductive system.
Question 3.9	Describe a typical nematode life-cycle.
Question 3.10	Name the different orders of nematodes.

PARASITE EXAMPLES GROUPED ACCORDING TO LIFE-CYCLE

4

Nearly every vertebrate is infected with a parasite at some stage of its life. Many of the parasites are host-specific. The parasites of man and domestic animals tend to be the most well known and the most widely distributed and are the parasites mainly investigated by parasitologists. In this section there are descriptions of some of these parasites, described in relation to the number of hosts that they infect during their life-cycle. The parasites using only one host during their life-cycle are considered to have a direct life-cycle; and those with more than one host are described as having indirect life-cycles.

■ 4.1 DIRECT LIFE-CYCLES

The parasites invade only one host during their life-cycle. This occurs in both protozoan and helminth parasites.

■ 4.1.1 *ENTEROBIUS VERMICULARIS*

A gut-dwelling nematode commonly known as the human pinworm. This parasite has worldwide distribution and almost all humans (mainly during childhood) are infected for a brief spell with this parasite.

- The adult females are 8–13 mm long; and the males are 2–5 mm long and 0.2–0.6 mm wide.
- Embryonated eggs — the infective stage — are swallowed and the larvae hatch in the gut and then migrate to the lower regions of the intestine, mainly the colon. The larvae undergo at least four moults before maturing into adults. Fertile females move down the colon and 'crawl' out of the anus onto the perianal region where they then deposit their eggs. The females usually die or return to the colon (see Fig. 4.1).

The most reliable method of diagnosis is to place a strip of cellotape along the anal region. Any eggs deposited will stick to the tape. The tape is placed on a microscope slide and then examined under a microscope for eggs.

The presence of worms and eggs around the anal region can be the cause of a certain amount of irritability and scratching. From the parasite's point of view this helps with dispersal. The eggs have a resistant shell and can withstand a certain amount of

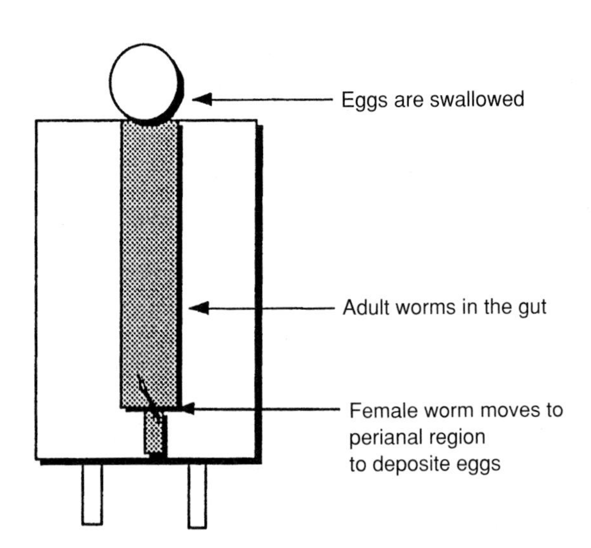

• **Figure 4.1** *Enterobius vermicularis*, a nematode commonly known as the human pinworm, has a direct life-cycle and lives in the lower regions of the alimentary canal. The females emerge out of the anus to deposit eggs around the anal opening causing an 'itchy bottom'.

desiccation. Apart from causing a 'itchiness' in the anal region and slight behavioural problems, the parasite has little or no pathological effect upon the host.

■ 4.1.2 *ASCARIS* SPP

This genus contains some the most commonly found large nematodes that inhabit the small intestine of vertebrates, especially mammals. *Ascaris lumbricoides* lives in humans and *A. suum* in pigs. Nearly all vertebrates have a specific species of *Ascaris*.

- The adult females of *A. lumbriciodes* can be up to 35 cm long and the males up to 30 cm; and both can be about 1 cm wide. Both live within the upper reaches of the small intestine.
- A mature female can produce over a 100,000 eggs per day which pass out of the body via the faeces. If the eggs are deposited in suitable soil they embryonate and remain until swallowed by the next host. If the eggs are deposited in a non-soil environment, they can survive for a long period (provided that they are not subject to continuous desiccation) and once back in the soil they embryonate.
- If the eggs are swallowed by a suitable host, the gut environment (pH, temperature, O_2 concentration, digestive juices etc) stimulates the larvae to hatch out.
- The first stage larvae moult into second stage larvae which penetrate the walls of the intestine (the mucosa) and enter into the lamina propria of the villi. The larvae penetrate into capillary blood vessels and then migrate to the liver via the portal blood vessels.
- From the liver the larvae travel via the pulmonary circulatory system and enter the lungs as third stage larvae. These migrate up the epiglottis and are swallowed and re-enter the small intestine where they moult twice before maturing into adult worms.

A relatively large number of larvae migrating at any one time can be the cause of a certain amount of pathology, known as *visceral larvae migrans*. The migrations through the lungs can lead to pneumonia-like symptoms. As the parasites move through the tissues they cause both local and general inflammation, partly due to their physical presence and partly to various secretions (mainly metabolic waste products).

A general increase in the number of eosinophils (eosinophilia) and the serum antibodies IgG and IgE is often symptomatic of this infection. Large numbers of mature adults can cause mechanical blockages of the bowels.

4.1.3 *ANCYLOSTOMA* SPP (HOOKWORMS)

The adult hookworms live in the mucosa of the small intestine. They become embedded within the villi and use their biting mouthparts to feed off blood. They are able to secrete an anti-coagulant to keep the blood flowing.

- The adult females are about 12 × 0.6 mm and the males are 10 × 0.45 mm. An adult female produces about 28,000 eggs per day which pass out of the body via the faeces.
- Once the eggs make contact with warm moist soil they embryonate. Within 48 h the first stage larvae (rhabditiform) hatch out and feed on soil bacteria and debris.
- After two further moults while in the soil they develop into the infective stage, the filariform larvae. These larvae crawl into a position, such as on blades of grass or a point high enough to make contact with humans.
- The filariform larvae can actively penetrate skin through hair follicles or damaged skin. Once they reach the dermal layers of the skin they first migrate along through the dermal layers and eventually enter a blood vessel.
- After moulting into fourth stage larvae they enter into the bronchi of the lungs. From the lungs they crawl up the throat and are swallowed and the fifth stage larvae emerge into the gut (see Fig. 4.2).
- Adult females produce eggs within about 40 days.

The migrations through the skin cause a skin irritation known as cutaneous larvae migrans. During the lung phase these parasites have an effect upon the host similar to *Ascaris*.

Ascaris and *Ancylostoma* are characterised by having larvae that migrate from the site of infection to the gut where they mature into adult worms. The eggs of *Ascaris*, once

• **Figure 4.2** A. *Ascaris* spp. A large intestinal nematode with a direct life-cycle. Larvae hatch out of the egg in the small intestine, burrow through the wall of the gut, migrate through the liver to the heart, lungs and back to the gut where they develop into adult worms. During the migrations the larvae undergo a series of moults. The migration process is known as *visceral larvae migrans*.
B. *Ancylostoma* spp. Nematodes commonly known as the hookworms. Infective stages are skin-penetrating larvae that survive in damp soil. Once the larvae have penetrated the skin, they migrate through dermal layers of the skin (*cutaneous larvae migrans*) and then via the blood to the heart, lungs and finally settle in the small intestine. The mouths of the adults are equipped with biting mouthparts. They secrete anti-coagulants and feed on blood from the intestinal wall. If the larvae are accidentally swallowed they can apparently avoid the migration.

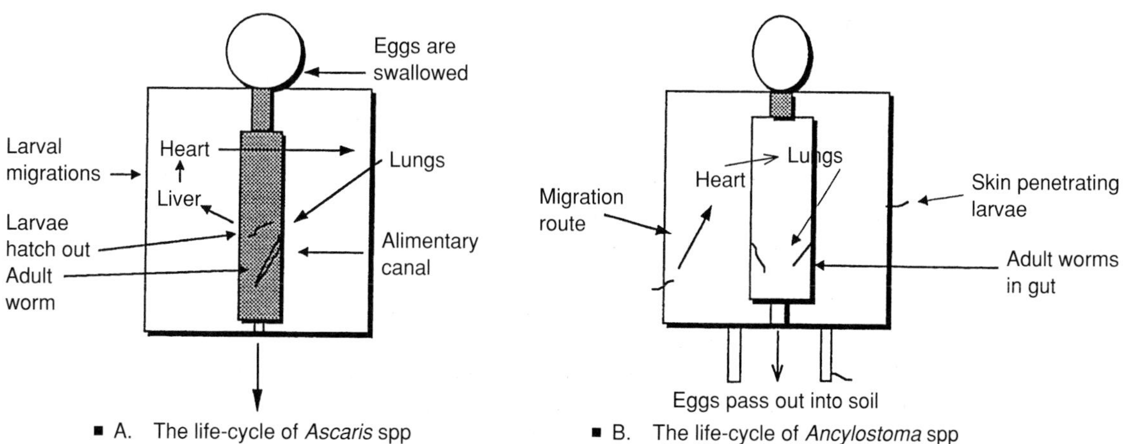

■ A. The life-cycle of *Ascaris* spp ■ B. The life-cycle of *Ancylostoma* spp

swallowed, end up in the gut where the larvae hatch out and then proceed to undergo the migratory phase. All ascarid larvae have a compulsory migratory route which starts where the adult parasite eventually ends up, ie the lumen of the ileum. It is interesting to speculate as to why the larvae appear to be compelled to undergo the migratory phase. If *Ancylostoma* larvae are accidentally swallowed instead of having to penetrate through the skin they remain in the gut and do not migrate round the body.

An explanation for the migration of ascarid larvae could be that each larval stage requires a different physiological and biochemical environment; and the different organs encountered en route represent intermediate hosts. Alternatively, migration could be an escape from the host's immune system.

4.1.4 *TRICHINELLA SPIRALIS*

The adult stages of this nematode are found in the gut and larval stages in striated muscle. The host of *T. spiralis* is first the definitive host and then the same host serves as an intermediate host (see Fig. 4.3).

- The adults live in the small intestine. The females are 3 mm long and the males 1.5 mm. The worms mature rapidly and copulation takes place after about 30 h. Female worms have been found to be inseminated after 36 h. Each female gives birth to about 1,000 live larvae.
- The newborn larvae burrow through the gut mucosa into the lamina propria and are then distributed around the body via the blood circulation. Further development of the larvae only occurs within striated muscle.
- After reproduction and the birth of live larvae the adult worms are eventually actively expelled from the gut.

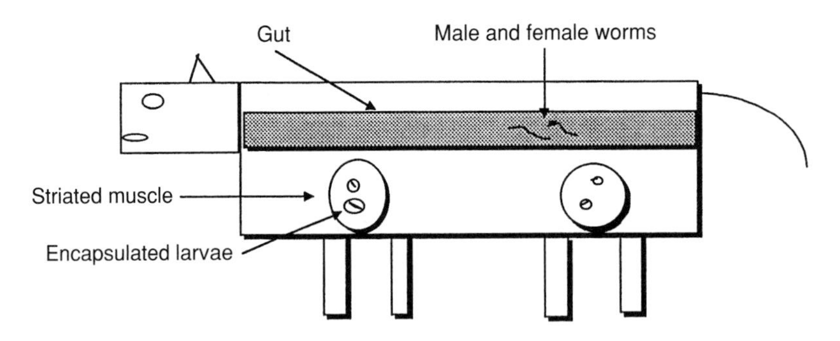

■ A. Life-cycle of *Trichinella spiralis*

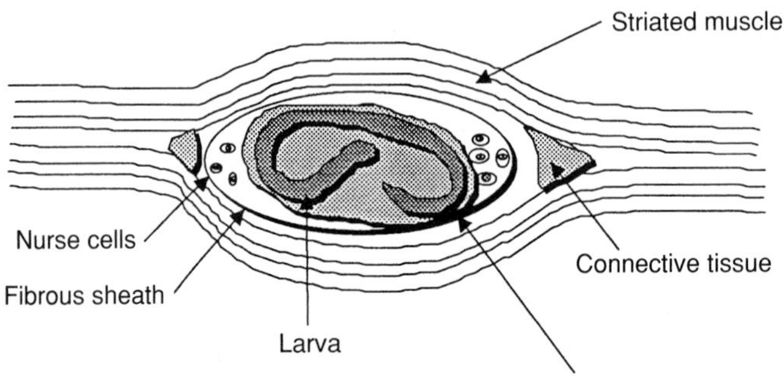

■ B. *Trichinella spiralis* Larva encapsulated in striated muscle

• **Figure 4.3** A. *Trichinella spiralis* is a widely distributed nematode that uses the same host as both an intermediate and definitive host. The mature male and female adults are found within the gut. The female gives birth to live larvae which then burrow through the gut wall and migrate to and encyst in striated muscle. The next larval phase develops when the muscles are eaten.
B. A larva of *T. spiralis* encysted within host striated muscle. The larva adopts a characteristic spiral shape and is surrounded by nurse cells enclosed within a fibrous sheath.

- The larvae that migrate into the striated muscles eventually become 'encysted'. The muscle cells that surround the parasite, the so-called nurse cells, provide the nutrients and stimulus for larval development. After 20 days a cyst composed mainly of host material with some input from the parasite has formed around each larva. The larvae coil into a characteristic 'spiral' shape within the cyst. As the cyst ages the host deposits calcium on the outer wall of the cyst. Calcified cysts can be detected by X-ray analysis.

The movement of the larvae from the gut lumen into the lamina propia causes local inflammation and infiltration of leukocytes. Those larvae that do not enter striated muscle often end up in the heart and central nervous system where they can cause disruption of the functioning of those particular organs.

The adults are eventually expelled from the gut whereas the larvae of *Trichinella* migrate to the striated muscle and remain within the host. Striated muscle from the parasite's point of view provides an ideal environment. Once within the muscle tissue the parasite is only exposed to limited aspects of the immune system but is in the ideal place for transference to the next host. Muscle tissue is nearly always the preferred food of carnivores.

4.1.5 *STRONGYLOIDES* SPP

Strongyloides stercoralis are small nematodes (2 × 0.04 mm) that live in the small intestine of humans.

- A mature worm first develops a male gonad and produces sperm. The male gonad then regresses and a functioning female gonad develops followed by self-fertilisation. Apparently parasitic males do not exist or have not yet been found.
- After fertilisation the embryonated eggs in the intestinal lumen hatch into first stage larvae. The larvae moult in the colon into second stage larvae and these are deposited, via the faeces, into the soil.
- Larvae that are delayed in the colon moult into third stage larvae, burrow into the gut mucosa and enter the circulation — a process known as autoinfection.
- Larvae that are deposited into relatively warm and damp soil moult twice more before become free-living adults.
- Adults of both sexes are found in the free-living phase and females produce embryonated eggs. The larvae hatching from these eggs moult four times before developing into free-living adults.
- Third stage free-living filariform larvae are infective and when the situation arises are able to penetrate the skin of a suitable host (see Fig. 4.4). The migration route of the larvae through the body to the gut is similar to that of *Ancylostoma*.

4.1.6 *TOXOCARA CANIS*

Toxocara canis is a zoonotic nematode parasite of man and dogs. The life-history of this parasite is to a certain extent controlled by the hormones of the adult female dog (see Fig. 4.5).

- Adult worms, found only in puppies, are 5–8 cm long. Eggs are only passed by puppies up to the age of five weeks. If the soil is oxygenated and warm the eggs take about 2–6 weeks to embryonate and become infective.
- A lactating bitch can become infected by ingesting eggs while cleaning the puppies and the subsequent larvae enter a dormant phase.

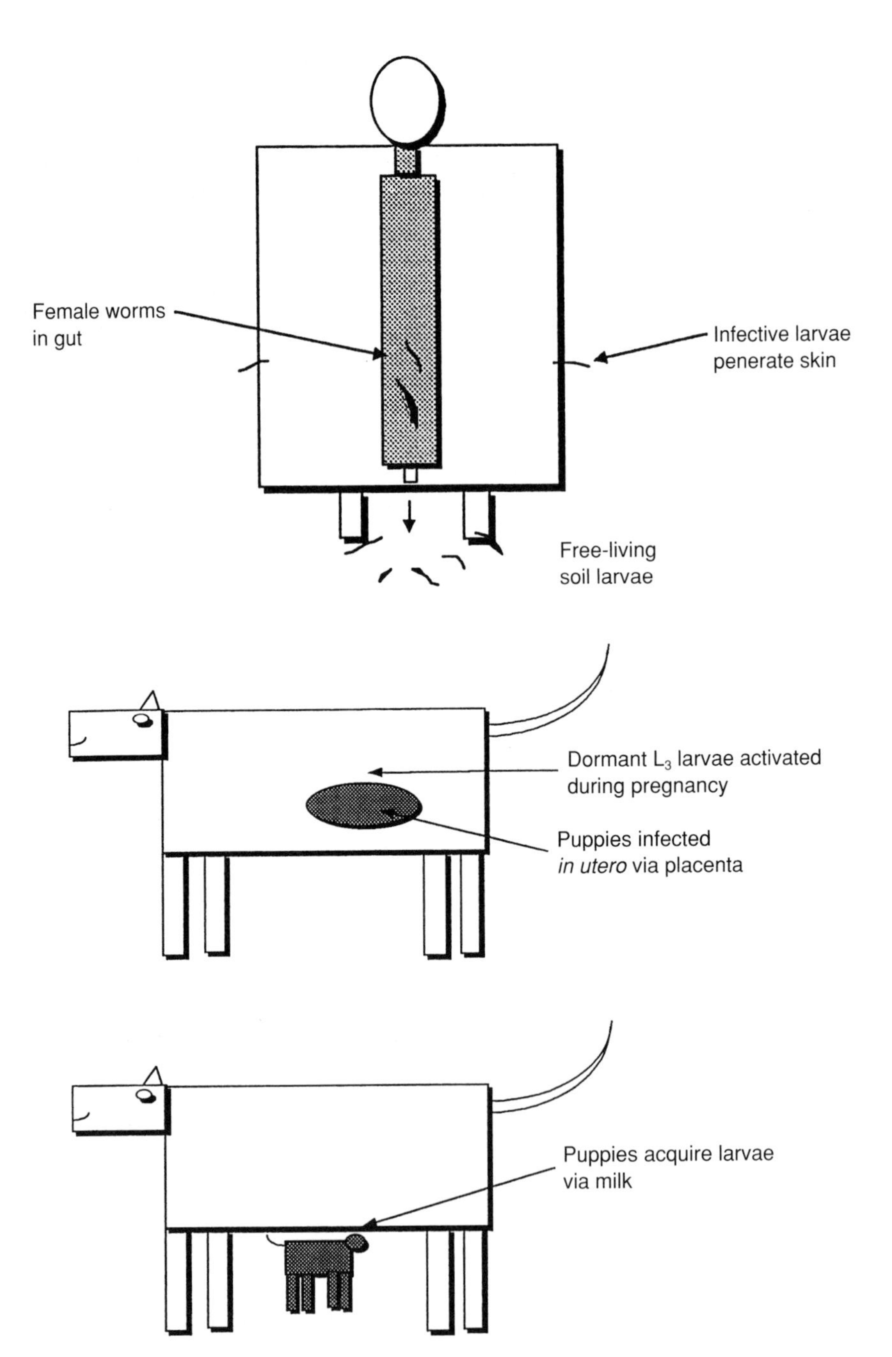

• **Figure 4.4** *Strongyloides stercoralis* is an intestinal nematode. Free-living soil larvae infect the host by penetrating through the skin. The larva migrates to the gut. The mature worm first develops a male gonad and after sperm production the female gonads form. There is self-fertilisation. The first stage larvae hatch out of the eggs in the gut lumen. Larvae pass out into the soil via the faeces. Single sex (homogonic) or male and female (heterogonic) free-living larvae develop in the soil.

• **Figure 4.5** *Toxocara canis* is a parasitic nematode found mainly in domestic dogs. Dormant third stage larvae are activated during pregnancy and the puppies are infected *in utero* and/or via milk. The worms mature in the pups and the mother can become re-infected from eggs in the pups' faeces.

- Adult worms live in the gut of puppies and bitch; eggs pass out via faeces; eggs containing L_2 larvae are ingested by dogs (accidentally by man); in bitch and puppies larvae migrate from gut to lungs and back to the gut.

- Embryonated eggs are swallowed and the larvae hatch into the gut lumen. Larvae (L_2) penetrate through the gut mucosa of the dog and then proceed to migrate to the lungs; and then eventually back to the gut mucosa to remain dormant within the tissues.

- Once the bitch becomes pregnant her hormones reactivate the larvae which then migrate into the foetuses via the placenta. Puppies can also become infected by larvae in the milk.
- Humans become infected by accidentally ingesting eggs. The L_2 larvae, now in the 'wrong host', burrow through the gut mucosa and then begin to migrate via the circulation round the host.
- If they migrate through the tissues the condition is known as *visceral larvae migrans* and if the larvae become trapped in the eye it is referred to as *ocular larvae migrans*.
- The larvae in man or rodents (referred to as paratenic hosts) do not develop further after completing their migration.

4.2 THE MONOGENEA: PARASITIC TREMATODES WITH ONLY ONE HOST

The monogenea are mostly ectoparasites of fish and amphibia and have only one host and, like nearly all trematodes, the adults are hermaphrodite. Examples include: *Diclidophora merlangi*, an ectoparasite found on the gills of the fish *Merlangius merlangus* (whiting); and *Diplozoon paradoxum*, an ectoparasite found on the gills of freshwater fish. The adult is comprised of two individuals that have become fused to each other.

4.2.1 *POLYSTOMA INTEGERRIUM*

The adult lives in the excretory bladder of a mature frog (*Rana temporaria*).

- A mature *Polystoma* is a flat worm, about 3 × 2 mm, with an adhesive organ called the haptor.
- The life-cycle of *P. integerrium* appears to be synchronised with that of its host. The parasite's genitalia are activated when the host prepares to enter the water in spring for mating. Large amount of eggs are released by the parasite once the frog is in the water. By the time the frog eggs hatch and reach the tadpole stage the *Polystoma* eggs hatch into a ciliated (five bands of cilia) oncomiracidium larva.
- The larva has a simple gut and a posterior sucker with 16 hooks. It enters the tadpole via the branchial pore and attaches to the host's gills.
- When the tadpole metamorphoses into a froglet, the juvenile larvae migrate over the ventral surface of the host to the cloaca and enter the bladder.
- Some of the larvae attach to the external gills of the tadpole and become neotenic, reaching sexual maturity within 20 days. These forms have only rudimentary copulatory organs, vaginae and vitellaria and a single testis. Cross-fertilisation occurs and the fertile eggs produce larvae which undergo normal development.

4.3 PARASITIC PROTOZOA WITH ONLY HOST

4.3.1 *EIMERIA* SPP

A protozoan intracellular parasite that lives in the epithelial cells of the gut mucosa (see Fig. 2.3).

- The infectious stage is the oocysts which are swallowed. The intestinal digestive juices dissolve the outer shell and release the sporozoites, which then invade the epithelial cells of the gut mucosa.
- Within each epithelial cell the process of schizogony commences.
- Schizogony is an asexual multiplicative phase in which the nucleus of the original sporozoite divides into several smaller nuclei and each daughter nucleus is surrounded by layer of cytoplasm producing numerous uninucleate merozoites.

- The merozoites burst out of the host cell and invade fresh epithelial cells. After several asexual multiplicative cycles, the merozoites transform into male and female gametes.
- The male gamete divides to form numerous flagellated microgametes.
- The female macrogamete develops from a single gametocyte and is fertilised by a microgamete. A wall forms around the fertilised zygote which develops into an oocyst. The nucleus of the oocyst divides into a uninucleate sporocyst (with a shell consisting of four walls) and each sporocyst divides into two sporozoites. The oocyst consists of four sporocysts, each containing two sporozoites which pass out of the body via the faeces.

4.3.2 *ENTAMOEBA HISTOLYTICA*

E. histolytica is a protozoan that lives predominantly in the lumen of the colon and caecum of humans.

- Resistant cysts are swallowed and an amoeba with four nuclei emerges from the cyst. Three mitotic divisions follow to form eight uninucleate amoeboid trophozoites. The trophozoites are capable of invading the cells of the intestinal wall (the epithelial cells) where most remain. However some enter the blood and then, via the circulation, invade the liver or the brain. Once inside the latter tissue, they are often the cause of amoebic abscesses.
- Trophozoites each with four nuclei form into cysts and pass out with the faeces.

4.3.3 *GIARDIA LAMBLIA*

A flagellate protozoan that lives on the surface of mammalian gut epithelial cells (see Figs 2.2 and 4.6).

- The active stage of this parasite is a binucleate, pear-shaped flagellated trophozoite. Each trophozoite has eight flagellae and a body 7–10 μm in diameter and 10–20 μm long.
- Two blepharoplasts (rigid structures) run along the length of the protozoan.
- The trophozoite attaches itself to the surface of the epithelial cells by a disc-like depression located at the anterior end.

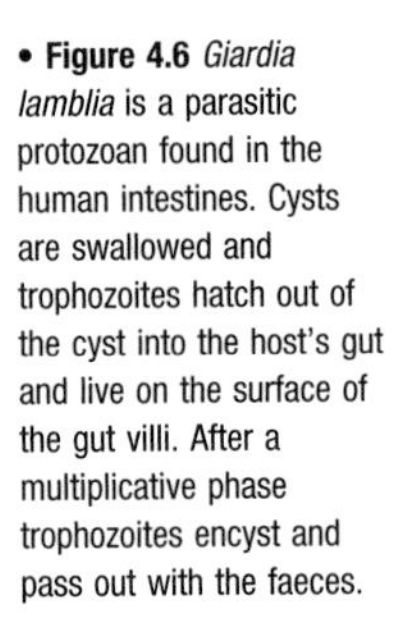

• **Figure 4.6** *Giardia lamblia* is a parasitic protozoan found in the human intestines. Cysts are swallowed and trophozoites hatch out of the cyst into the host's gut and live on the surface of the gut villi. After a multiplicative phase trophozoites encyst and pass out with the faeces.

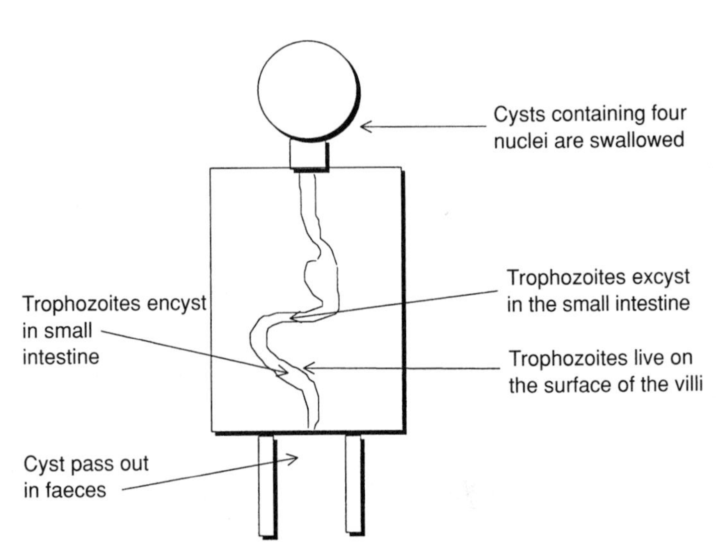

- The resting stage is an oval cyst with four nuclei that lacks certain subcellular organelles (mainly hydrogenosomes) and has attenuated flagellae and blepharoblasts. The cyst passes out of the host via the faeces and becomes the infective stage. The rigid outer wall protects this stage from environmental changes.
- Cysts, which are the infective stage, are swallowed and once within the gut two fully formed trophozoites grow within the cytoplasm. These become attached to the epithelial cells of the gut and then undergo the process of binary fission producing numerous trophozoites.

4.4 INDIRECT LIFE-CYCLES

These are parasites that have more than one host in their life-cycle. This group includes parasites that during the process of transfering from the definitive host to the intermediate host may have free-living stages. Alternatively they may use a vector to transmit the parasite from one host to another, thus avoiding the free-living distributive phase.

4.4.1 PROTOZOA WITH MORE THAN ONE HOST

4.4.1.1 *Plasmodium* spp

These are the intracellular blood parasites (see Figs 2.4 and 2.5) that cause the disease known as malaria. Four different species of *Plasmodium* infect humans and they are *P. falciparum*, *P. vivax*, *P. ovale* and *P. malariae*. The genus *Plasmodium* tends to be host-specific. The majority of vertebrates are infected either by species of *Plasmodium* or by a related intracellular protozoan parasite.

The transmission of *Plasmodium* from one host to another is via a blood sucking arthropod. In man it is the female mosquito of the Anophelinae or Culicinae that is responsible for the transmission of the parasites. Although there nearly 800 species of *Anopheles* and *Culex* only a limited number transmit malaria.

The mouthparts of the mosquito can pierce the skin of the host and reach into the capillaries to obtain a blood meal. If sporozoites — spindle shaped motile forms of the parasite — are present in the salivary glands of the mosquito, they are inoculated into the host while the mosquito absorbs blood from the host.

The sporozoites circulate round the body via the blood and within 1–30 min penetrate the liver and then into the hepatocytes — the exoerythrocytic (non-red blood cell) stage. Once established within an hepatocyte, the sporozoite undergoes a multiplicative process known as exoerythrocytic schizogony. Numerous uninucleate merozoites are formed as a result of schizogonony.

P. vivax sporozoites take 6–8 days to mature and produce about 10,000 merozoites; *P. ovale* takes 9 days to produce about 15,000 merozoites; *P. malariae* takes 12–16 days to produce 2,000 merozoites and *P. falciparum* 5–7 days to produce 40,000 merozoites, all from a single sporozoite.

The hepatic merozoites exit from the liver tissue into the circulating blood. They become attached to receptor sites located on red blood cells before invasion to form the erythrocytic stage. Once inside an erythrocyte the merozoite develops into a feeding trophozoite. A parasitophorous vacuole forms round the trophozoite and the trophozoite takes on a 'signet-ring' shape.

The trophozoite develops into a schizont which undergoes schizogony to produce numerous blood stage merozoites. The erythrocyte eventually bursts releasing the merozoites into the circulation and these then invade non-infected red blood cells. The result is a dramatic increase in parasitaemia and, when the number of parasites reaches a certain level, the process of schizogony and the subsequent liberation of merozoites becomes synchronised and periodic.

P. malariae has a 72 h periodic cycle whereas *P. ovale*, *P. vivax* and *P. facilparum* have a 48 h cycle. At the end of each cycle when the erythrocytes burst and release the merozoites into the bloodstream, 'toxins' are released into the host. Each time there is the liberation of parasites into the bloodstream it has a pyrogenic effect upon the host, that is the host suffers from fever.

- Once the merozoite population reaches a certain density some of the merozoites develop into male and female gametocytes. These tend to accumulate in the peripheral circulation and are taken up by the mosquito as it feeds.
- In the gut of the mosquito the male gametocyte produces numerous microgametes which are released to fertilise the larger female macrogamete. Once fertilised the female oocyte is known as a zygote.
- The zygote transforms into a motile ookinete which penetrates the gut lining of the mosquito and is associated with the outer wall of the gut. The ookinete now settles and develops into an oocyst. The nucleus of the oocyst divides and gives rise to numerous sporozoites which eventually make their way toward the mosquito's salivary gland.

4.4.1.2 *Trypanosoma spp*

The trypanosomes are parasitic flagellated protozoa referred to as kinetoplastid flagellates and are transmitted by insect vectors. Each individual has a single flagellum attached to a kinetoplast located within the cytoplasm. A kinetoplast is an organelle associated with part of the mitochondria and has its own DNA. The typical trypanosome form is elongate with a single flagellum.

The most well known genera are *Trypanosoma*, *Leishmania* and *Endotrypanum*. These parasites infect both man and domestic animals and many are zoonotic. A feature of this group of trypanosomes is that no sexual reproduction has yet been observed. Asexual multiplication occurs in both hosts.

There are two groups of trypanosomes that infect man: the Salivarian; and the Stercorarian trypanosomes.

- Salivarian eg *Trypanosoma brucei*. The vector host is *Glossina* (tsetse-fly) and while within the fly, the parasite develops in the midgut and then migrates to the salivary gland and is injected into the mammalian host when the fly feeds on blood (an inoculative infection).
- Stercorarian eg *T. cruzi*. These develop in the hindgut of the arthropod vector and the infective stages — the epimastigotes — pass out via the faeces onto the host skin while the insect extracts blood. The epimastigotes are 'rubbed' into the wound made by the fly's mouthparts (a contaminative infection).

4.4.1.3 *Trypanosoma brucei*

There are several subspecies of *T. brucei* such as *T. brucei gambiense* and *T. brucei rhodesiense* which are the main causes of African sleeping sickness (African trypanosomiasis).

- The fly injects the metacyclic (or metatrypanosome) form through the host's skin where it rapidly enters a blood vessel. Once in the bloodstream they transform into a trypanomastigote. These forms move by using their flagellum and an undulating membrane. This stage remains extracellular in the blood and body fluids and multiplies asexually by binary fission.

- After numerous divisions the parasites can reach a concentration of $1,500/mm^3$ and at this stage they begin to invade cerebrospinal fluid and interstitial spaces.
- At a certain population density a number of the trypanosomes transform into stumpy forms (an intermediate form). These tend to congregate in the peripheral blood vessels where they are taken up by the fly while it feeds.
- The stumpy forms migrate into the fly's midgut and transform into procyclic trypanomastigotes. They divide asexually and the new generation migrates to the salivary gland and proventriculus of the fly where they transform into the metacyclic form (metatrypanosomes). Depending upon the environmental temperature and the species of *Glossina*, the cycle within the fly takes about 25–50 days.

4.4.1.4 *Trypanosma cruzi*

T. cruzi is found in Central and S. America and the main vector hosts are the blood-feeding triatomid bugs *Rhodnius* spp and *Panstrongylus* spp. One of the main features of this parasite is that it is mostly intracellular.

- The vector bites a mammal near its eyes or mouth and defaecates at the site of the bite, which causes a local itch. The resultant scratching helps the infective stage of *T. cruzi* to enter the host at the site of the bite.
- The infective stage, the trypanomastigote, actively penetrates a host cell and transforms into an amastigote. Amastigotes undergo asexual multiplication forming a new generation of trypanomastigotes to be released into the bloodstream once the cell dies.
- The trypanomastigotes are able to invade a range of cells including cells of the reticuloendothelial system, heart muscle cells, glial cells and cells of the urinogenital tract.
- After several multiplication phases a new generation of amastigotes develops and accumulates in a primary chagoma, a lesion that forms on the face. When a vector feeds, it acquires from these sites amastigotes which transform into epimastigotes within the hindgut of the insect. These undergo repeated divisions and can remain within the bug for 1–2 years, ie the rest of the vector's life-span.

4.4.1.5 *Leishmania* spp

Leishmania spp are spread by the biting sandflies of the genera *Phlebotoma* and *Lutzomyia*. *L. tropica*, *L. donovani*, *L. mexicana*, *L. brasiliensis*, and *L. peruvana* are all species which infect humans. *Leishmania* spp are found throughout the warmer regions of the world and have numerous reservoir hosts (a zoonosis). *Leishmania* spp can cause either a cutaneous, mucocutaneous or visceral disease. The promastigotes are found within the gut of the vector and the amastigotes within the macrophages of the mammalian host.

- A feeding sandfly takes up blood containing amastigotes inside a macrophage. The insect digests the macrophages and releases the amastigotes into its stomach. These are transformed into promastigotes and undergo binary fission. The new generation of promastigotes migrates into the insect's proboscis ready to be injected into the next host during the insect's next meal.
- The promastigotes injected into the host are phagocytosed by the host's macrophages.
- The promastigotes are able to resist being lysed by the phagocytic cells and transform into amastigotes which undergo asexual division. The host cell is eventually disrupted, releasing the amastigotes into blood and body fluids.

- Lesions form in the host's skin close to the site of the insect bite and most infected cells remain in the vicinity of the lesion.

L. mexicana and *L. brasiliensis* are both capable of invading mucocartilage and mucous membranes and disfiguring lesions can develop (cutaneous leishmaniasis).

L. donovani promastigotes invade macrophages and transform into amastigotes which can invade macrophages within the spleen, liver and bone marrow (visceral leishmaniasis).

4.4.1.6 *Toxoplasma gondii*

Toxoplasma gondii is a protozoan coccidian parasite that has both an asexual and a sexual stage in its life-history. The definitive host is the cat (*Felis domestica*) and the intermediate host is a rodent.

- The infection in the rodent consists of a pseudocyst located in either muscle or brain. Each pseudocyst contains numerous bradyzoites which are released into the lumen of the cat's stomach. The individual bradyzoites invade the epithelial cells of the small intestine and undergo asexual multiplication to form merozoites.
- The released merozoites invade non-infected epithelial cells and the population of the parasites increases within the cat. Some of the merozoites develop into male and female gametes which fuse to form a zygote that transforms into an oocyte.
- The oocyte nucleus divides and numerous sporozoites from within the oocyst pass out via the cat's faeces.

The oocysts are eaten by rodents and accidentally by man. The oocyst releases the sporozoites into the gut of the rodent or man and these then invade intestinal macrophages.

- They are able to resist being digested by the macrophage and transform into tachyzoites. Binary fission occurs within the macrophage and daughter tachyzoites are released to infect other cells within the host.
- Once the population has reached a certain density the division of tachyzoites stops and they transform into pseudocysts within various tissues, especially the brain. This change in development may be due to the host's immune response.

If the host is a pregnant female the tachyzoites can cross the placenta and infect the foetus. In humans the tachyzoites invade the neural tissue, the myocardium, lungs and liver. Inflammatory lesions develop as a result of the cell invasion and this can lead to permanent damage in the foetus.

4.4.1.7 *Cryptosporidium* spp

Cryptosporidium is a gut protozoan parasite of man and domestic animals.

- Sporulated oocysts each containing four sporozoites are swallowed.
- The sporozoites emerge in the small intestine and become attached to the surface of the epithelial cells of the gut mucosa.
- The sporozoites are surrounded by elongated microvilli which fuse and envelop the sporozoite; and the parasite becomes intracellular and forms into a trophozoite.
- A schizont develops from the trophozoite which divides to form eight merozoites (first generation). The cell ruptures and releases the merozoites and these attach to new non-infected epithelial cells and by schizogony produce four second generation merozoites.

- The new merozoites are released and attach to a non-infected epithelial cell before becoming internalised. Within the new cells the merozoites transform into either a male microgametocyte or a female macrogametocyte. The male microgametocyte undergoes several divisions to produce microgametes which then fuse with the macrogametocyte to form a zygote.
- The zygote differentiates into an oocyst which is shed with the faeces. If the oocyst is swallowed by a new host they then sporulate.

4.5 PARASITIC HELMINTHS TRANSMITTED BY VECTORS

The majority of these are nematodes that do not have a free-living stage and rely on arthropods that feed on vertebrate body fluids for transmission from one host to another.

4.5.1 *WUCHERERIA BANCROFTI*

W. bancrofti is a filarian nematode that is one of the causes of filariasis in humans. As far as is known this parasite is specific to humans and has no reservoir hosts. There are three 'geographical variants' each with a different genus of mosquito as its vector. In the South Pacific the vector is a species of *Aedes*, a mosquito that feeds nocturnally.

The adult *W. bancrofti* worms live within the lymphatic system of the lower limbs of humans.

- The male is approximately 4 × 0.1 cm and the female 8 × 0.3 cm long. The females are ovaviviparous and give birth to live microfilariae (L_1 larvae) which enter the lymphatic system and then into the circulating bloodstream.
- The microfilariae require a relatively high oxygen tension and tend to concentrate in the capillaries of the lungs during the day. During the night while the host is sleeping the oxygen tension of the peripheral blood increases and the microfilariae migrate out of the lungs. This coincides with the nocturnal feeding habits of the mosquito. During feeding the mosquito takes up microfilariae with the blood meal.
- The larvae first accumulate within the flight muscles of the mosquito and then undergo three moults to form L_3 larvae. These larvae migrate to the mosquito's mouthparts and are deposited on the skin while feeding. The larvae 'crawl' through the wound and migrate through the subcutaneous tissues into the lymphatic system and take about a year to reach maturity. After copulation the females give birth to live microfilariae, a process which can continue for up to 8 years.

4.5.2 *ONCHOCERCA VOLVULUS*

The adult parasite (a nematode) lives in the subcutaneous tissues of (mainly) humans and the microfilariae migrate round the body. If a larva reaches the host's eye and remains there this condition can lead to permanent blindness, commonly known as river blindness.

The vector host is the biting blackfly *Simulium damnosum* which completes its life-cycle in the 'white water' of river rapids.

- From the subcutaneous tissues microfilariae enter the fly while it feeds. The microfilariae move from the fly's mouth into the muscle fibres of the fly's thorax. After two moults the larvae migrate into the proboscis to be in position for the next occasion when the fly feeds. During feeding the larvae withdraw from the proboscis. The larvae are deposited on the skin and enter the host through the wound.

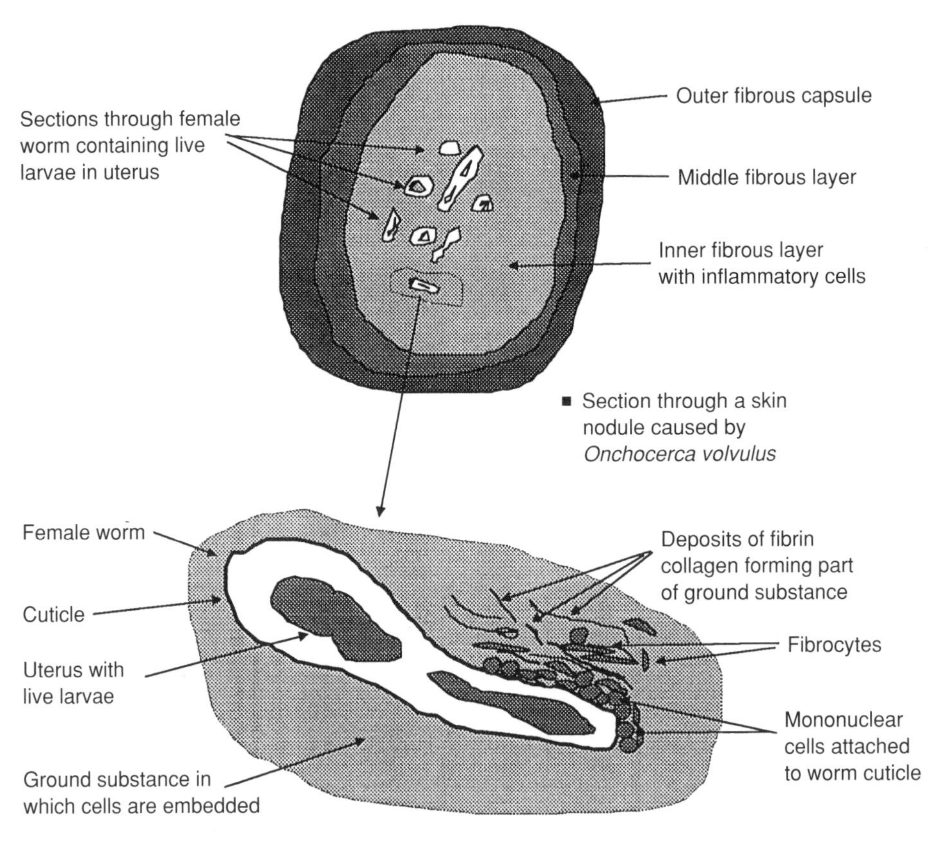

• **Figure 4.7** The adults of the nematode *Onchocerca* spp live in the subcutaneous tissues of mammals. Old established worms become enclosed in skin nodules. Each nodule normally contains a small male and a large female worm and numerous larvae in the nodular tissue.

- Development of the worm takes place within the subcutaneous tissues of the human host. The adult females can grow to a length of 40 × 0.3 cm whereas the males are by comparison very small, 0.3 × 0.15 cm. In young infections the worms migrate through the subcutaneous tissue. In older established infections the worms slow and eventually cease moving and coil up within a subcutaneous nodule (see Fig. 4.7). Normally a nodule contains a single male and female. The female continues to produce larvae many of which remain just below the surface of the skin.

Brugia malayi, *Loa loa*, *Dipetalonema perstans*, and *Mansonella ozzardi* are further examples of filarian nematodes which have arthropod vectors.

4.5.3 *LOA LOA*

The adult worm (a nematode) migrates through the subcutaneous tissues of its human host. The mature female is 60 × 0.5 mm and the male 32 × 0.4 mm. The female produces live microfilariae which penetrate capillaries and then circulate round the body via the bloodstream.

- A small blood sucking fly (*Chrysops*) pierces the skin of the host and the microfilariae are taken up with a blood meal. The larvae first reside in fat bodies inside the stomach of the fly. The L_3 larvae, the infective stage, emerges within 10 days and migrates to the biting mouthparts.

- The infective stage larvae are released into the wound as the fly feeds. Once in the host's subcutaneous tissues they moult and develop into adults and remain in the host for up to 4 years.

4.5.4 *DRACUNCULUS MEDINENSIS*

D. medinensis (a nematode) is distributed throughout India, the Middle East, Central Africa and S. America. It is a zoonotic parasite and apart from man can live in monkeys, horses, dogs, raccoons, foxes and cattle. The intermediate host is a freshwater copepod, *Cyclops* spp, *Mesocyclops* spp and *Thermocyclops* spp.

- The adult worms usually live in the lower limbs of the definitive host. The female is the larger with a length of about 100 cm × 1.5 mm and the male is 40 × 0.4 mm.
- The female in the lower limbs positions itself with the oral cavity facing downwards. Around the mouth region the host tissue becomes ulcerated. As the female matures the uterus undergoes a 'prolapse' with the result that the uterine opening lies close to the oral cavity. An ulcer forms on the host's skin around the anterior end of the worm. When the limb comes into contact with water the ulcer 'bursts' and motile larvae are released from the uterus into the oral cavity and these then escape into the water.
- If the larvae are eaten by a copepod they develop, after about 3 weeks, in the copepod's haemocoel into an infective L_3 larva.
- When the copepod is swallowed by the definitive host the larvae are released into the lumen of the host's intestine. The larvae migrate through the gut wall into the connective tissues and eventually to the subcutaneous tissues of the limbs. The migration takes about a year and during this period there are two further moults.

4.6 PARASITES WITH TWO HOSTS AND FREE-LIVING DISTRIBUTIVE PHASES

All of the Eucestoda (with a few exceptions) use two hosts to complete their life-cycle.

4.6.1 *TAENIA* SPP

Taenia saginata (the beef tapeworm) — also known as *Taeniarhynchus saginatus* — and *Taenia solium* (the pork tapeworm) adults normally live only in the human gut. Both parasites are long-lived and have a worldwide distribution. *T. saginata* can grow to the length of 10 m and *T. solium* 6 m and both are normally found within the small intestine.

- A scolex of a mature adult becomes embedded within the gut mucosa and is anchored by four suckers. *T. solium* has a ring of hooks, the rostellum, at the tip of the scolex. *T. saginata* does not have a rostellum. Behind the scolex is the 'neck' region from which the proglottids originate. The youngest proglottid is nearest to the neck region and the oldest furthest away. Each proglottid contains a set of male and female organs.
- It has been speculated that cross-fertilisation occurs between different proglottids. The gravid proglottids, ie those containing only fertilised eggs, are shed into the gut lumen and pass out via the faeces. In some cases the proglottids are shed as 'ribbons' out of the anus. Some shed proglottids can 'wriggle' across the grass where they are ingested by the intermediate host.
- If the proglottids are eaten, the embryonated eggs are released in the gut of the intermediate host. Alternatively the eggs are swallowed via the vegetation.
- The shelled eggs contain a hexacanth embryo (the oncosphere). The gut environment stimulates the oncosphere to hatch out of the egg. The oncosphere is motile and

migrates from the gut mucosa into and through the epithelium to be distributed round the body. Skeletal muscle is the common destination for the larva and once on or within the muscle, it develops into a cysticercus (bladderworm), the metacestode stage.

- A cysticercus is a round fluid-filled cyst and folded in from the outer tegument into the fluid is an invaginated scolex. The cysticercus remains in an inactivated state within the muscle until eaten by the definitive host. The scolex evaginates in the gut of the definitive host and attaches to the gut epithelium; and development of the adult stage then commences.

4.6.2 *DIPHYLLOBOTHRIUM LATUM*

The adult tapeworm lives in the gut of humans as well as fish-eating carnivores. In humans the adult worm can reach a length of 10 m.

- The scolex has two attachment organs called bothria and no hooks.
- Mature proglottids have a centrally located uterine pore and the eggs are shed into the gut lumen and pass out with the faeces. If the eggs reach fresh water, a free-swimming coracidium larva hatches out.
- The coracidium is eaten by the first intermediate host *Diaptomas* (a small crustacean) and develops into the first metacestode stage — the procercoid — in the body cavity of *Diaptomas*.
- The crustacean is eaten by a fish — the second intermediate host — and the procercoid burrows through the intestinal wall and into the muscle where it develops into a plerocercoid — the second metacestode stage.
- The fish is eaten by man or a fish-eating mammal and the plerocercoid matures in the small intestine into an adult worm. Should the fish be eaten by another predator fish the plerocercoid remains in the muscle of the predator fish which simply acts as a carrier host.

4.7 CESTODES WITH AN ASEXUAL MULTIPLICATIVE PHASE IN THE INTERMEDIATE HOST

4.7.1 *ECHINOCCUS GRANULOSUS*

In contrast to the tapeworms already described the adult of *E. granulosus* is the smallest tapeworm (see Fig. 4.8). The metacestode stage is known as a hydatid cyst and when fully developed is one of the largest of the metacestodes.

- The definitive host is usually a carnivore, with the domestic dog the most frequently detected host. The adult tapeworm consists of a scolex, with at most three proglottids and is only 5 mm long. An infected dog can harbour 1,000+ adult worms. However the total biomass is probably less than one adult *T. saginata*.
- After fertilisation the end proglottid becomes filled with embryonated eggs (gravid). It then detaches and fertilised eggs are released into the gut lumen which escape from host via the faeces.
- Man and several different types of herbivore are potential intermediate hosts. The eggs are swallowed and the digestion processes of the host help the oncosphere to hatch.
- The oncosphere moves into the gut mucosa and then penetrates through the gut epithelium into the lamina propria. Once inside the host's tissue the larva migrates via the blood circulation until it reaches one of the larger organs, such as the liver

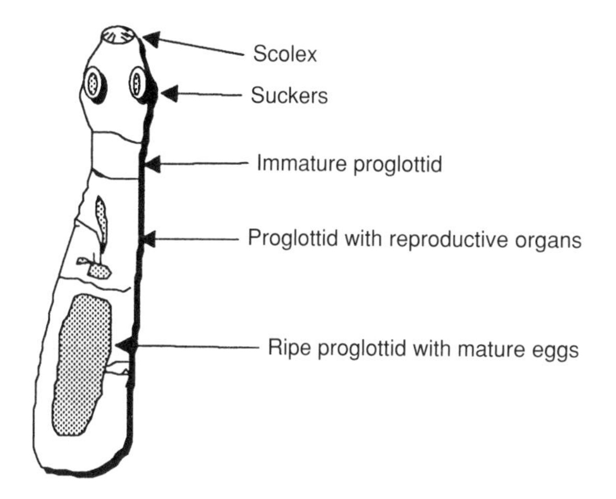

• **Figure 4.8** The adult stage of the tapeworm *Echinoccocus granulosus* is the smallest of the tapeworms and consists of a scolex and three proglottids. The adult worms are parasites of domestic dogs and other canine species as well as some of the larger carnivores such as lions.

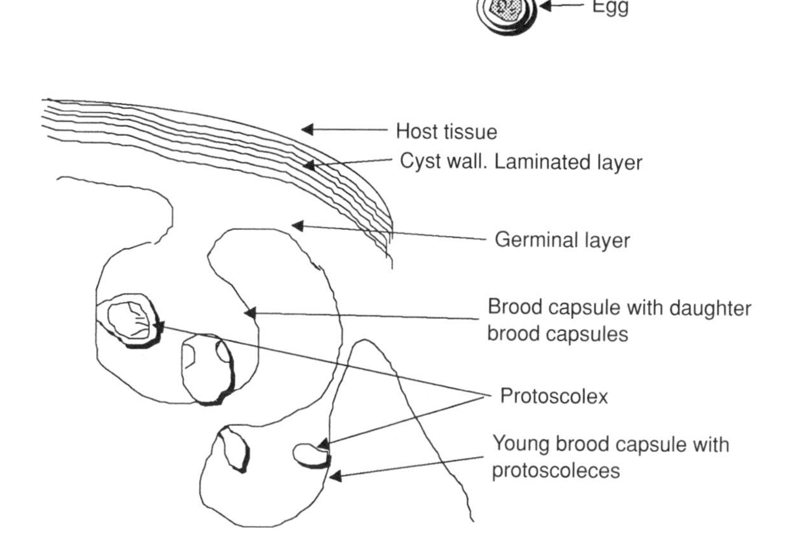

• **Figure 4.9** The hydatid cyst of *E. granulosus* is the largest of the tapeworm metacestodes. It is a large fluid-filled cyst and can measure up to 10 cm in diameter. The innermost layer of the cyst is the fertile membrane from which arises the brood capsule containing protoscoleces.

or lungs, where it settles and develops into a hydatid cyst (the metacestode stage, see Fig. 4.9).

- As soon as the oncosphere has become attached to tissue, it secretes a hyaline membrane around itself. The membrane thickens and differentiates into an inner germinal layer and an outer acellular layer.
- The germinal layer produces protoscoleces as well as more of the laminate layer. A tissue layer surrounds the protoscolices to form a brood capsule. The brood capsules, each containing several protoscoleces, detach from the germinal membrane and remain with the fluid-filled hydatid cyst. The hydatid cyst continues to grow and reaches a size of up to 10 cm in diameter. Each brood capsule if released from the cyst is capable of forming secondary cysts.
- If the intermediate host plus the hydatid cyst is eaten by the definitive host the protoscoleces are released and each has the potential to develop into the adult tapeworm.

4.7.2 *TAENIA CRASSICEPS*

The adult *T. crassciceps* lives in the gut of a fox (*Vulpes vulpes*). It has a scolex with four suckers and a rostellum with a double row of hooks. The adult is 12–20 cm long.

- The mature proglottids produce eggs which pass out with the faeces and are eaten by a rodent.
- The oncosphere hatches out of the egg and then penetrates through the epithelial cells into the peritoneal cavity of the rodent intermediate host.
- In the peritoneal cavity it develops into a cysticercus characteristic of the taeniids. Each cysticercus (the metacestode stage) has an invaginated scolex with four suckers and a rostellum and is capable of exogenous budding. Each bud develops into a cyst containing an invaginated scolex. These cysts detach from the mother cyst and then repeat the budding process. This can continue for the rest of the life of the intermediate host.

4.7.3 *MESOCESTOIDES CORTI*

The adult *Mesocestoides corti* parasitises the gut of carnivores. The adult has four suckers (acetabula) but no rostellum and is about 4–8 cm long. The complete life-cycle of this parasite is not yet known.

- Eggs pass out via the faeces which are thought to be eaten by an orbatid mite, the first intermediate host.
- The second intermediate host is a rodent or other small vertebrate including lizards (*Lacertes* spp).
- Once the mite is eaten the metacestode, known as a tetrathyridium, develops in the second intermediate host. The tetrathyridium is normally a solid structure but can become cystic if conditions change.
- Each tetrathyridium has four acetabula and a rostellum and can produce an exogenous bud which grows almost to the size of the original before separating and this gives the appearance of having divided by longitudinal fission.
- Adult worms apparently also have the capability of asexual budding.

4.8 HELMINTH PARASITES WITH AN INTERMEDIATE HOST AND TWO FREE-LIVING STAGES

4.8.1 *SCHISTOSOMA* SPP

The Schistosomatidae are the one family of trematodes in which the sexes are separate. The adult male is 10 mm long and the female 15 mm but much narrower. When both are mature the female becomes permanently associated with the male which lives within her gynocophoric groove.

- There are four species of *Schistosoma* that are infective to man and they are *S. mansoni*, *S. japonicum*, *S. haematobium* and *S. intercalatum*. All of these can cause schistosomiasis, a disease more commonly known as bilharzia.
- *S. mansoni* has the most widespread distribution and also can be maintained in laboratory animals.
- The four different species are identified by the external appearance of the eggs. Except for *S. japonicum* the eggs have a single spine but the position of the spine differs in each of the other three species.

The tegument of the adult worm differs in each species. *S. mansoni* has a tegument covered with papillae, in *S. haematobium* the papillae are shorter and more widely spaced and *S. japonicum* has a smooth tegument.

The intermediate hosts are freshwater snails and each species of worm has a different genus or species of snail host. *S. mansoni* uses the snail host *Biomphalaria* spp; *S. japonicum* has *Oncomelania* spp and both *S. haematobium* and *S. intercalatum* have *Bulinus*.

S. mansoni adults live in the inferior mesenteric blood vessels, *S. japonicum* in the superior mesenteric blood vessels and *S. haematobium* adults are found in the venous plexus associated with the urinary bladder.

- Although trematodes have a mouth and gut they also absorb food through their tegument and most of it will be amino-acids and sugar associated with blood.
- Both male and female worms have suckers which are used for attachment. In the female the birth pore is above the posterior sucker which can penetrate the endothelium cells of the walls of the blood vessels.
- The *S. mansoni*, *S. japonicum* and *S. intercalatum* eggs enter the connective tissue through damaged blood vessels. They use the spine on the exterior shell as well as proteolytic enzymes secreted by the embryonic miracidium within the egg to burrow through the sub-mucosa of the gut into the gut lumen and then escape to the exterior via the faeces.

As many as 50% of the eggs of the above species do not exit from the host but via the circulation end up in the liver where they remain and eventually the host reaction forms a granuloma around the egg. *S. haematobium* eggs burrow through the blood vessels into the bladder and pass out of the host via the urine. Eggs that do not pass out of the host become lodged in the bladder wall and urinary tracts and granulomas form round them.

- From the eggs that pass out of the host into fresh water with temperatures of about 25–28°C and average amount of light, a free-living ciliated miracidium hatches (see Fig. 4.10).
- The miracidia remain active for 6–8 h, and swim until they come across a specific species of aquatic snail. Once they encounter the snail they actively penetrate the foot of the snail and migrate to the snail's digestive organ, the hepatopancreas.
- The germ cells develop into sporocysts, which can produce a second generation of sporocysts as well as the next generation, the cercaria.
- Cercariae exit from the snail and are free-living. They have a tail to help them propel themselves through the water to seek out their next and definitive host. Each cercaria is either male or female and can survive in water for about 8 h.
- Once a cercaria comes in contact with mammalian skin (in particular human skin), they actively penetrate via a hair follicle, shed their tail and enter the dermis, a process which can take up to 2 h. In the dermis they transform into juveniles known as schistosomules and about 2 days later enter the bloodstream and migrate to the lungs (the lung stage worms). From the lungs they migrate to the liver and eventually to mesenteric or bladder wall blood vessels where maturation takes place. Eggs are usually produced after about 28 days.

4.8.2 *FASCIOLA HEPATICA*

Fasciola hepatica is commonly known as the liver fluke and, like the schistosomes, is a digenean trematode with two hosts and two free-living stages (see Fig. 4.11). This parasite

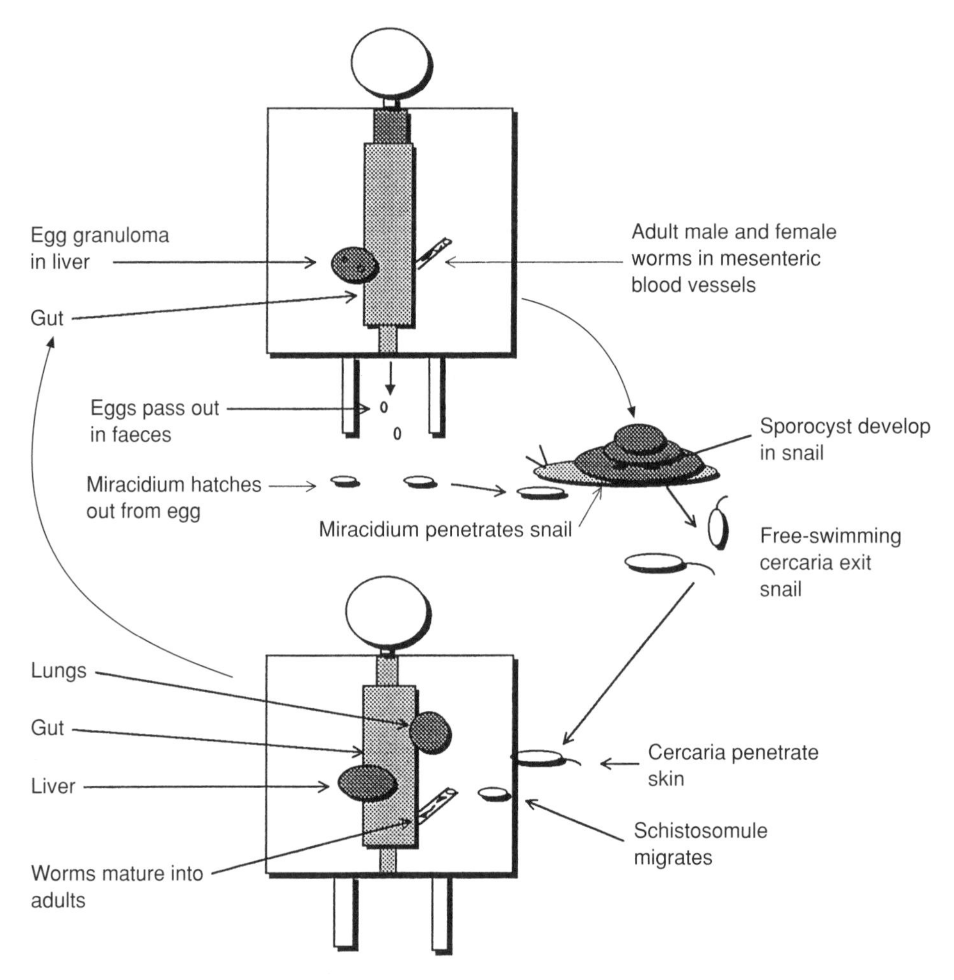

• **Figure 4.10** The adult males and females of the trematode *Schistosoma mansoni* live within the mesenteric blood vessels. The eggs pass out with the faeces and hatch in fresh water into a free-living miracidium. The miracidium invades a fresh water snail and develops into a sporocyst. Within the sporocyst, cercariae with forked tails develop. They escape into fresh water from the snail and swim until they come in contact with the skin of the host. The cercariae penetrate the skin, shed their tails, and the juveniles (schistosomulae) enter the dermal layers and slowly migrate via the lungs to liver and the mesenteric blood vessels. It takes about 28 days to complete the migration and mature.

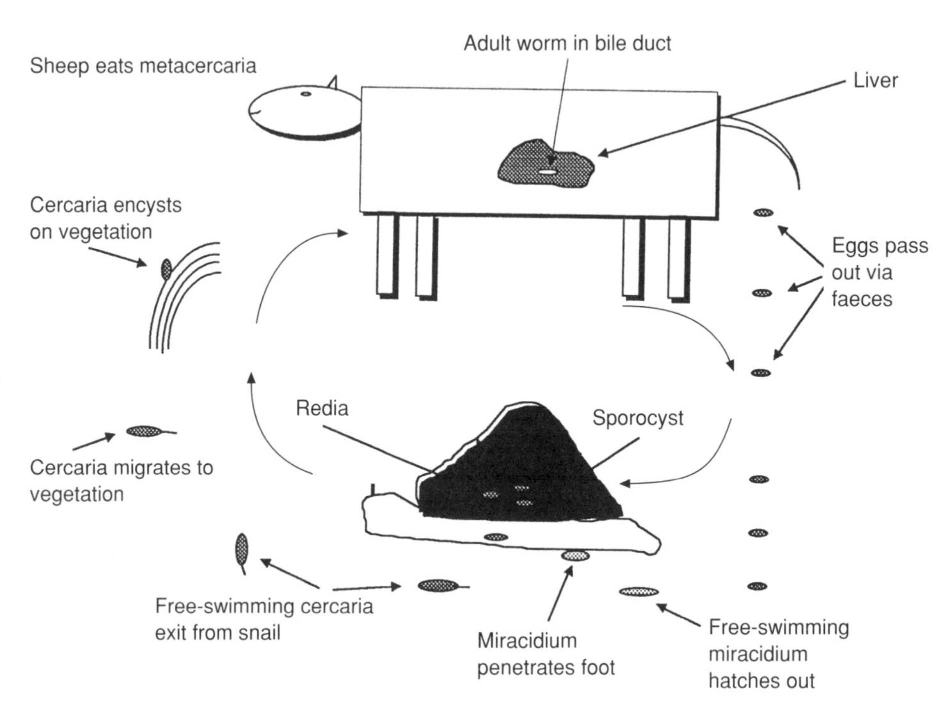

• **Figure 4.11** *Fasciola hepatica* is commonly known as the liver fluke. The adult lives in the bile duct. The eggs pass out via the faeces into fresh water and hatch into a miracidium. The miracidia invade an aquatic snail. Within the snail the miracidium forms first into a sporocyst, then into a redia and finally into a cercaria. The free-swimming cercaria encysts on vegetation and remains until eaten.

is found wherever there is intensive sheep farming and relatively damp conditions. Sheep is the dominant definitive host but man and other herbivores can also become infected. The adults, like the majority of trematodes, are hermaphrodite.

- The adult flukes inhabit the bile duct and have a flattened leaf-like shape. They grow to about 35×15 mm. A mature individual produces eggs which pass into the bile duct, then into the intestine and finally out of the host with the faeces. If the eggs are deposited in fresh water, a free-swimming miracidium hatches out after 9 days. The eggs have an operculum (a lid-like structure) through which the miracidium emerges.
- The ciliated miracidium swims until it comes across *Lymnaea* spp — a freshwater snail — and penetrates through the snail's soft tissues into the hepatopancreas. Once settled, the germ cells within the miracidium develop into sporocysts and usually produce a second generation of sporocysts. From the germ cells within the sporocysts the next generation forms — the redia. Within the redia the germ cells develop into cercariae. The cercariae exit from the redia and eventually from the snail and settle on grass. Once on a blade of grass they encyst and are now referred to as metacercariae.
- If a metacercaria is eaten the cercariae emerge from their cyst penetrate the wall of the host's gut and grow into juvenile flukes. The juveniles migrate through the abdominal cavity over the visceral organs until they reach the liver. They actively penetrate the liver and migrate through the hepatic tissues until they reach the bile duct.

■ SUMMARY

Endoparasites either have a direct life-cycle, that is involving only one host, or they have an indirect life-cycle, ie more than one host. Almost all the parasites have a sexual mutliplicative phase (exceptions are some of the zooflagellates) which occurs in the definitive host. In addition many parasites have an asexual multiplicative phase that may take place in either the definitive or the intermediate host.

The distributive phases can occur when the parasite passes out of the host a larva or is transmitted by an arthropod vector or if the intermediate host is eaten by the definitive host.

END OF CHAPTER QUESTIONS

Question 4.1 Name the different phases that can occur during a the life-cycle of a parasite.

Question 4.2 What aspect of the parasite's life-cycle determines the description of the host?

Question 4.3 What is the difference between direct and indirect life-cycle?

Question 4.4 What does 'internal accumulation' mean with regard to parasite life-cycles?

Question 4.5 Give examples of parasites that have only one host and no asexual reproduction.

Question 4.6 Name examples of parasites that reproduce asexually; and for which no sexual reproductive phase has been observed.

Question 4.7 Which group of parasites have both asexual and sexual reproduction within the same host?

Question 4.8 Using named examples explain the terms homogonic and heterogonic.

Question 4.9 Name parasites that have two hosts but no multiplicative phase in the intermediate host.

Question 4.10 What developments take place in the above type of intermediate host?

Question 4.11 Explain how larval stages that develop in intermediate hosts are transmitted to the definitive hosts.

Question 4.12 Give an example of a parasite that uses the same host as both definitive and intermediate host.

Question 4.13 Give examples of parasites that have two hosts and multiplicative phases in both hosts.

Question 4.14 Discuss with examples the relevance of multiplicative phases that occur in the intermediate hosts.

Question 4.15 Describe the life-cycle of a parasite involving three hosts and describe what type of developments occur in the second intermediate host.

INTRODUCTION TO HOST RESPONSE

5

Each vertebrate host is a multicellular organism composed of tens of millions of cells. Most cells form tissues and organs but each cell has at some stage of its life a nucleus which contains a set of chromosomes specific to each organism — the genotype. The genes (made up of DNA) are located on the chromosomes and contain the 'blueprint' for all the protein molecules that are assembled into an individual.

Every cell type has an outer surface membrane, the structure of which is unique. Cells identify/recognise each other by their surface membranes, ie they can distinguish self from non-self. A normal healthy individual will attempt to maintain its integrity by eliminating any foreign (non-self) cells. An invading parasite fits this category, that is as far as the host is concerned a parasite, no matter how well adapted it is to the host or how tolerant the host is of the parasite, is identified by the host's surveillance mechanism as non-self or foreign.

In order to maintain integrity and keep out or destroy infectious parasites, vertebrates have evolved a protective system. This system is based upon:

- having a protective outer layer — the skin — which maintains a barrier to all other non-self organisms.
- having a mechanism for detecting any non-self organism that has managed to breach this protective layer. This system is based upon having cells whose prime function is detection and destruction of non-self material (in a primary infection) and then 'remembering' the infectious material, to be able react more rapidly during a secondary infection.

The mechanism that operates against a primary infection, in a host with no previous contact with the parasite, reacts in a similar manner no matter what the infectious organism may be and is referred to as the innate immune system.

If the parasite is not destroyed by the innate immune response and survives to become a chronic infection, then a complex adaptive immune response comes into operation. This mechanism leads to the host developing a memory of the infection and becoming programmed to react much more rapidly during a secondary infection.

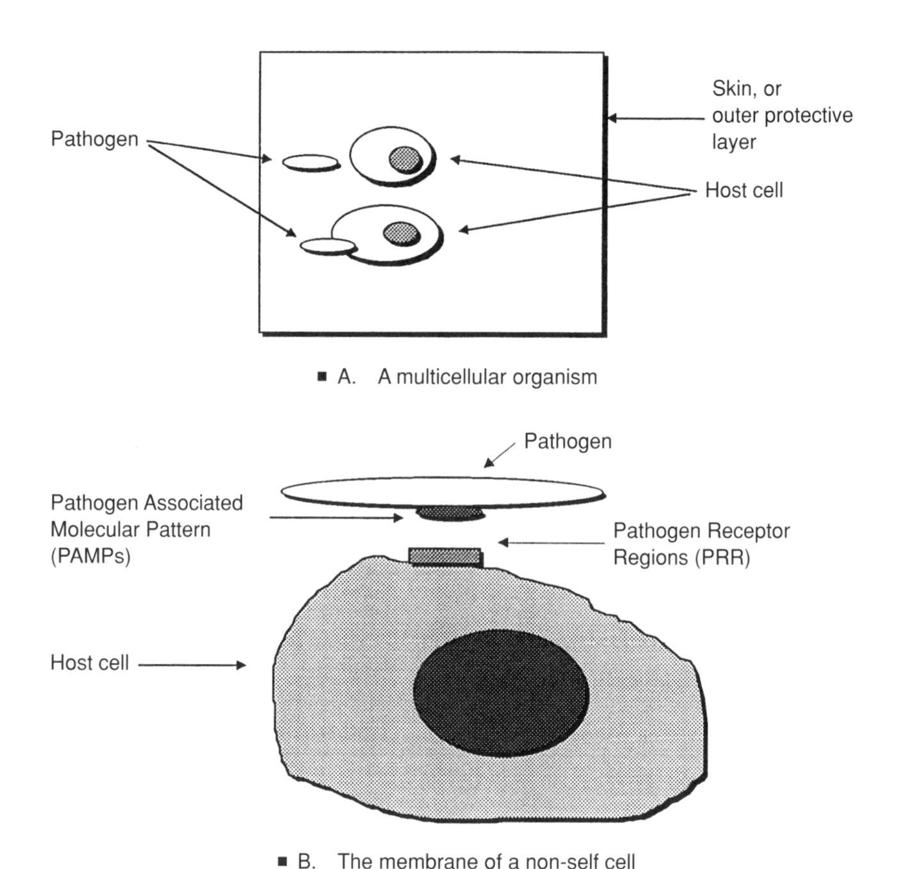

• **Figure 5.1** A. In a multicellular organism all the host cells have identical outer membranes, whereas the membrane of a pathogen is derived from a different genotype and is different.
B. The membrane of a non-self cell/organism or pathogen is different from that of the host cells. The host surveillance cells recognise this difference. On the surface membrane of the pathogen are molecules known as Pathogen Associated Molecular Patterns (PAMPs). Host surveillance cells have on their membranes groups of molecules known as Pathogen Receptor Regions (PRR) which can detect PAMPs.

■ 5.1 INNATE IMMUNITY

The innate immune system probably came into existence once multicellular organisms evolved. As soon as the division of labour between cells or specialisation of cells became operative, phagocytic cells acted aggressively toward non-self material that had managed to penetrate through the outer covering or 'skin'. As organisms became more complex so did protective mechanisms (see Fig. 5.1).

Firstly there is the skin barrier which produces secretions containing bactericides and fungicides to prevent other organisms from living on it or trying pass through it. Secondly a whole range of wandering phagocytic cells and natural killer (NK) cells have evolved to destroy any organism that breaches the skin barrier.

In all animals the innate immune system predates the adaptive immune system and the reasons for believing that are:

- Innate immune systems are found in all multicellular organisms, whereas adaptive immune systems are found only in vertebrates.
- Innate immune recognition distinguishes self from non-self perfectly, a condition not completely met by the adaptive immune response.
- The innate immune system uses receptors that can be traced back to invertebrates, whereas adaptive immunity appears to use mechanisms that lead to clonally specific antibodies and T cell antigen receptors (TCRs).

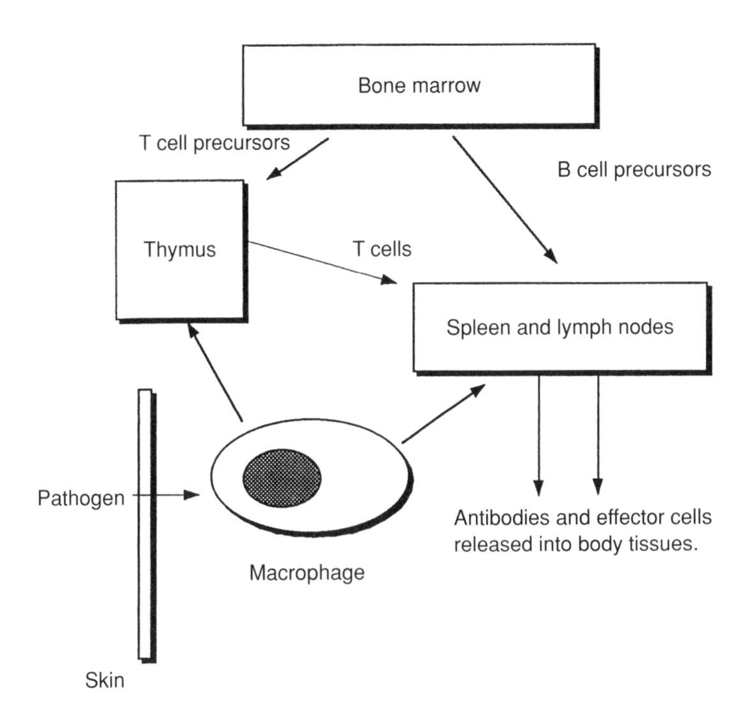

• **Figure 5.2** Once a pathogen enters the body through the skin (or through the cells lining the alimentary canal or the lungs) a phagocytic surveillance cell (eg macrophage or dendritic cell) engulfs the pathogen. The macrophage then functions as an antigen-presenting cell (APC). The APC migrates to the lymphatic organs to present the antigen to T cells and thus start an adaptive immune response.

However and wherever a pathogen gains entry into the tissues, it will be attacked by one of the many phagocytic cells that are normally present. If the pathogen is not completely eliminated, specialised phagocytic cells known as antigen-presenting cells (APCs) will convey 'extracts' of the pathogen (known as antigen) to the lymphatic system.

It is in the lymphatic organs such as the thymus spleen and lymph nodes where the antigen 'reacts' with the lymphocytes, which is the beginning of the adaptive immune response (see Fig. 5.2).

The problem with recognising pathogens in general is their molecular variability and in some instances their high mutational rate which alters their surface structure. However, pathogenic helminth and protozoan parasites tend to be less variable than parasitic micro-organisms and viruses. They seem to be able to resist the initial innate immune response and survive for considerably longer periods within the host. If that occurs then the infection becomes chronic and the immune reaction develops into an adaptive immune response.

The mechanisms that come into operation, that is the recognition of the parasite by the phagocytic cells and subsequent antigen presentation to the lymphocytes, are thought to have their origins in the primitive systems that still may operate in certain invertebrates. If the current mammalian innate immune system is derived from the original phagocytic cells which recognise molecular structures of pathogen membranes, this then implies that there must be a finite number of these molecular patterns. So the surveillance cells of vertebrates, having evolved from the invertebrates, can probably recognise the molecular patterns associated with the external membranes of most pathogens. These considerations help the innate immune system to recognise and attempt to destroy the target (the pathogen).

- The major requirement is that the operative cells of the innate immune system can distinguish between self and non-self.
- The most important aspects of the innate immune system are that it is the first line of defence against any infectious agent and that it is non-specific.

5.2 ENTRY OF A PATHOGEN

Skin consists of two basic layers: the outer epidermis and the inner dermis. The dermis is composed mainly of connective tissue — a reticulum of fibroblasts embedded in a semi-fluid ground substance — with blood vessels, nerve endings and various cells including mast cells. The composition of the dermis is normally stable. The equilibrium is chemical as well as physical and any change is first detected locally. Damage to the skin such as a cut or abrasion provides rapid access to the dermal layers for pathogens.

Alterations to the local skin environment can result in changes to the local temperature, pH and oxygen concentration. All, or any of these changes can affect the stability of the local cells, especially the mast cells.

The endothelial cells that line the walls of capillary blood vessels have membrane-bound proteins which allow one cell to react with another and these are known as intercellular adhesion molecules (IAM). These molecules bind to a specific group of molecules — the ligand located on a cell membrane such as a neutrophil. Each adhesion molecule is capable of binding to more than one ligand using different binding sites.

In general the binding affinity to individual adhesion molecules is low but due to the abundance of these molecules and ligands there is a high interaction. The interaction between the endothelium and circulating leukocytes involves a number of different pairs of these molecules. The function of the adhesion molecules is to control the aggregation and migration of leukocytes into sites of infection.

Most of the leukocytes (neutrophils, monocytes, eosinophils, lymphocytes etc,) express ligands on their membranes and these then bind to matching adhesion molecules. Soon after any skin or tissue damage, the adhesion molecules and their ligands are activated. This leads to the binding of circulating leukocytes to the endothelium.

Destabilisation of the local equilibrium can lead to mast cell degranulation. Mast cells are granular cells which contain the precursors to numerous pharmacological

BOX 5.1

The different types of adhesion molecules are:

- *Selectins* — transmembrane molecules such as endothelial adhesion molecules (ELAM-1) which have extracellular domains that bind to the carbohydrate residues present on the appropriate ligands.
- The *immunoglobulin* super gene family is expressed on the vascular endothelium and includes intercellular adhesion molecules (ICAM-1, ICAM-2) and vascular cellular adhesion molecule (VCAM).
- *Integrins*, involved in the interactions of platelets and neutrophils at inflammatory sites or sites of vascular damage, leukocyte adhesion to immune cells and binding of cells to the extracellular matrix.
- *Vascular addressins* expressed on endothelial venules and on the endothelium at sites of chronic inflammation. These molecules modulate lymphocyte traffic into secondary lymphoid tissue and inflammatory sites.

compounds, most notably histamine. The products of the degranulated mast cells, mainly histamine, heparin, bradykinin, 5-HT (5-hydroxytryptamine) etc, affect the permeability of cell membranes particularly the endothelial cells and also 'loosen' the binding of these cells to one another. This enables the leukocytes that have congregated at the site where the adhesion molecules have been activated to migrate into the tissues where the tissue damage or pathogen invasion has occurred. Once the cells have passed out of the lumen of the blood vessel, they interact with collagen, laminin, fibronectin, the extracellular matrix and tissue cells.

This has two effects, firstly fluid leaks out from the capillary lumen through the endothelial cells into the damaged tissues and secondly the end-on walls between the endothelial cells loosen, which permits leukocytes to migrate into the tissues. The local infected tissue becomes swollen with the increase in fluid.

- Neutrophils — granulocyte leukocytes — are the first cells to migrate from the capillary lumen to the damaged tissues and will attempt to phagocytose any invading pathogens, assisting the local resident phagocytic cells. The neutrophils release neurotoxins into the infected area which cause the sensation of local pain.
- The mast cells also release serotonin and bradykinin and slow release substances which add to general swelling and pain.
- The local area becomes swollen and red with the increase of fluid, in particular blood, and there is also pain all of which are symptoms of inflammation (see section 7.2).

5.3 THE HOST'S RESPONSE TO THE PATHOGEN

The majority of helminth and protozoan parasites are able to resist being phagocytosed by neutrophils. However neutrophils are relatively short-lived and are replaced by eosinophils and blood monocytes. If the parasite is first damaged and then destroyed, fibroblasts invade the damaged area and the deposited fibrocytes bind together and secrete collagen and fibrinogen to form scar tissue. During the healing process dead cells and cellular debris are discarded in the form of a cellular exudate (pus).

If the parasite is not destroyed and does not develop into a life-threatening disease, the infection then changes from the acute to the chronic phase during which the parasite can either migrate and/or multiply or remain. If the parasite migrates it could move to a preferred site or become systemic. The parasites migrate by moving through the ground substance matrix in the connective tissues or via the body fluids, mainly the bloodstream.

Once the resident phagocytic cells (the surveillance cells) have identified an invading pathogen by the methods just described, their function is to destroy it. Once the pathogen associated molecular patterns (PAMPs) on the surface membrane of the parasite are recognised by the pathogen receptors regions (PRRs) on the surveillance cells, this induces the secretion of cytokines (inflammatory cytokines) by the surveillance cells.

Cytokines are proteins secreted by cells prominent in the immune response eg T cells and macrophages that act like hormones affecting the behaviour of similar or different cell phenotypes. If the cytokine stimulates self cells it is referred to as being autocrine but if the cytokine stimulates different cells it is known as paracrine. The cytokines act on specific cell receptors.

On a cell surface membrane there are cell surface molecules which are identified by specific monoclonal antibodies and referred to as CD (Cluster of Differentiation) molecules or markers.

- The inflammatory cytokines such as interleukin-1 (IL-1), tumour necrosis factor-alpha (TNF-α), IL-6 and interferon-gamma (IFN-γ) mediate an inflammatory response.
- Chemical signals derived from PAMPs (pathogen associated molecular patterns) initiate the activation of a series of membrane-bound molecules known as costimulators. These molecules assist in the activation of T lymphocyte cells (see section 5.6.2). The costimulators on the T cells are CD28 and on the antigen-presenting cells they are B7 molecules.
- The effector functions of specified cytokines (eg IL-4, IL-5, IL-10, IL-12, IFN-γ) and transitional growth factor beta (TGF-β) are controlled by signals derived from the PAMPs.
- Parasites can induce the secretion of inflammatory cytokines that lead to the stimulation of the cytotoxic cells of the innate immune system.
- IL-1 and TNF-α (inflammatory cytokines) activate adhesion molecules on endothelial cells and/or stimulate chemotaxis and hence induce the migration of antigen-stimulated lymphocytes to the site of infection.
- IL-6 (an inflammatory cytokine) acts on activated B cells, converting them into immunoglobulin secreting plasma cells (this becomes a part of the adaptive immune response).

If the parasite is destroyed by a combination of inflammatory products such as reactive nitrogen or oxygen (radicals) and a cellular attack, the innate immune system has completed its function. However, if the parasite survives or is only partially destroyed, the process of the innate immune system continues.

5.4 THE ONSET OF THE SPECIFIC IMMUNE RESPONSE

If the neutrophils are not effective at killing the pathogen they are replaced by eosinophils, monocytes, lymphocytes, fibroblasts and variety of other phagocytic cells and generally have no role in the specific immune response.

The monocytes that migrate from the blood into the tissues are now known as tissue histiocytes. These cells, together with resident macrophages and phagocytic cells, initiate the specific immune response. The phagocytic macrophages that can identify the infectious agent, such as a parasite, then attempt to engulf it in an 'amoeboid' manner. Many parasites have the ability to resist phagocytosis. There are protozoans such as *Leishmania* that actually invade macrophages and survive within them.

However, phagocytes appear to ingest secretions or fragments of most parasites, particularly helminths. Once the parasitic material is absorbed into the cytoplasm of the macrophage, it is digested by lysosomes into peptide fragments (see Fig. 5.3). These peptide fragments (antigen peptides) are 'moved' through the cell cytoplasm to the surface membrane where they become associated with the major histocompatibility complex (MHC) molecules attached to the surface membrane. There are three classes of MHC molecules, the most important of which are the class I and class II protein molecules.

The macrophages that express the antigen peptides associated with the MHC molecules are referred to as antigen-presenting cells (APCs). These antigen primed cells migrate to the nearest lymph node or the spleen (see Fig. 5.4).

The APCs end up in the cortex and paracortex regions of the lymph nodes or spleen. The role of the APC is now to present the antigen peptide fragment to the lymphocytes, in particular the T lymphocytes (T cells).

After the first contact with a pathogen the now stimulated macrophages secrete a range of cytokines and likewise the T cells release cytokines when in contact with the

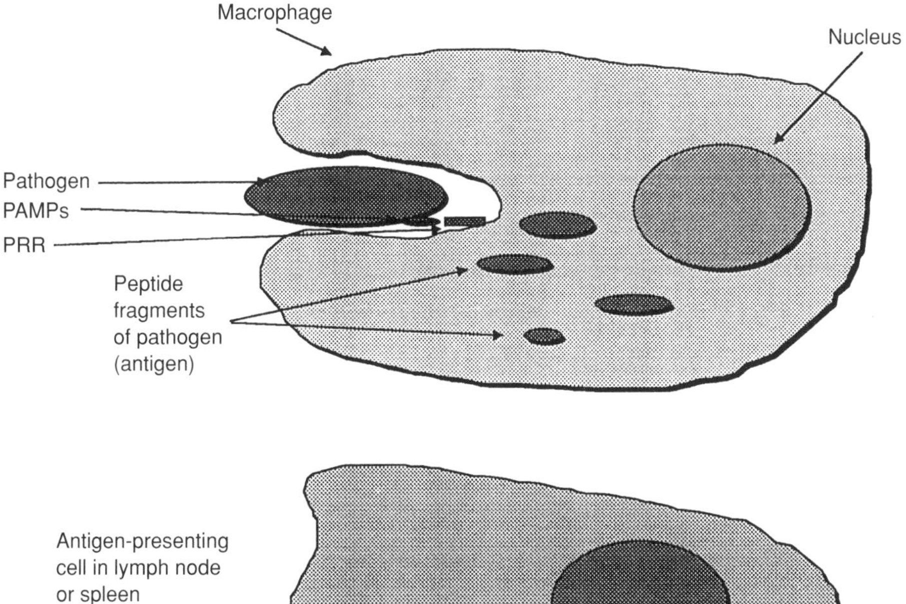

• **Figure 5.3** The Pathogen Receptor Region (PRR) on phagocytic surveillance cells such as macrophages identifies the Pathogen Associate Molecular Pattern (PAMP) on the pathogen. The phagocytic cell engulfs the pathogen. Once inside the cell it is digested and peptide fragments of the pathogen are released.

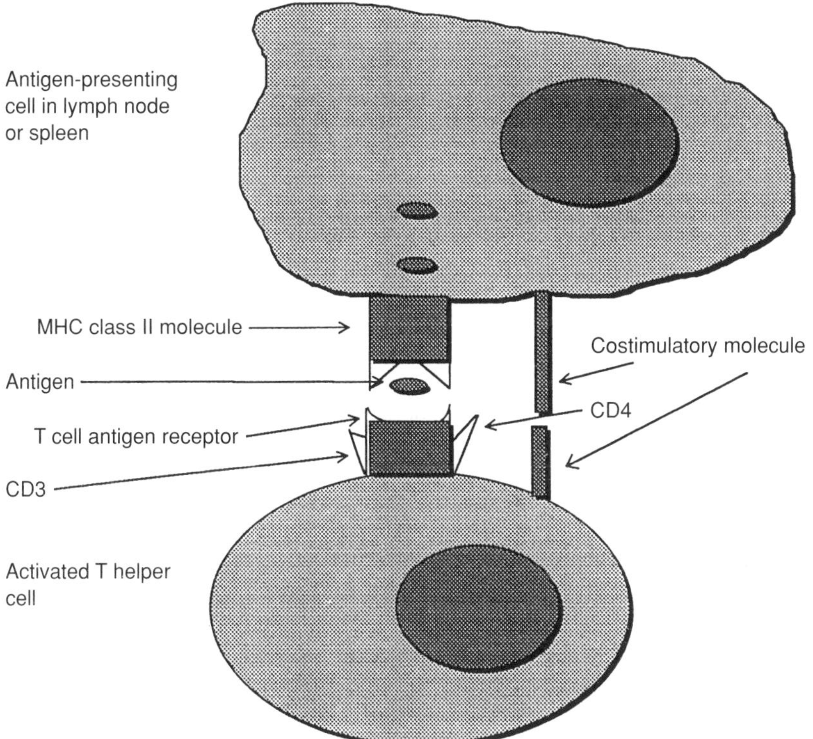

• **Figure 5.4** Within a phagocyte the digested peptide fragments of the pathogen are transported to the cell surface where they become associated with the Major Histocompatibility Complex (MHC) class II molecules. The costimulatory molecules on the phagocyte are activated. The phagocytic cell now referred to as Antigen Presenting Cell (APC) and migrates to a lymph node or spleen where it presents the antigen peptide to a T helper cell.

APC. In addition and occurring simultaneously, a whole range of wandering phagocytic cells and natural killer (NK) cells become primed. Circulating soluble compounds such as complement, cytokines and soluble antibodies are now released into the system.

■ 5.5 AN OUTLINE OF THE ADAPTIVE IMMUNE SYSTEM

The organs that are involved in the immune response are bone marrow, thymus, spleen and the numerous lymph nodes scattered round the body, including those associated with the gut, lungs etc. Bone marrow and the thymus are the primary lymphoid organs and the others are secondary lymphoid organs.

The lymphatic system is linked up via the blood circulatory system and a series of lymphatic vessels which circulate lymph fluid. The lymphocyte is the main cell associated with the lymphatic system and has a dominant role in the development of the adaptive immune response. Nevertheless all white blood cells (leukocytes), phagocytic cells, mast cells and keratinocytes have significant roles in the immune system.

5.5.1 THE LYMPHOID ORGANS

Bone marrow is a primary lymphoid organ where all the cells of the immune system originate (see Fig. 5.5) and where the B lymphocytes (B cells) are processed (see Fig. 5.6). The cells of the immune system are derived from basic all purpose (pluripotent) stem cells located within the red bone marrow in the central cavity of bones. In normal healthy adults the most active red bone marrow is in the ribs and sternum.

The red bone marrow has two zones: the haemopoietic (blood forming — consisting of sinusoids surrounding a central vein) and vascular zones. The sinusoids contain the progenitor or blast cells such as myeloblasts, erythroblasts and pluripotent stem cells. (Myeloblasts give rise to granulocytes; erythroblasts give rise to red blood cells and pluripotent stem cells (PSCs) develop into lymphoid cells and platelets.)

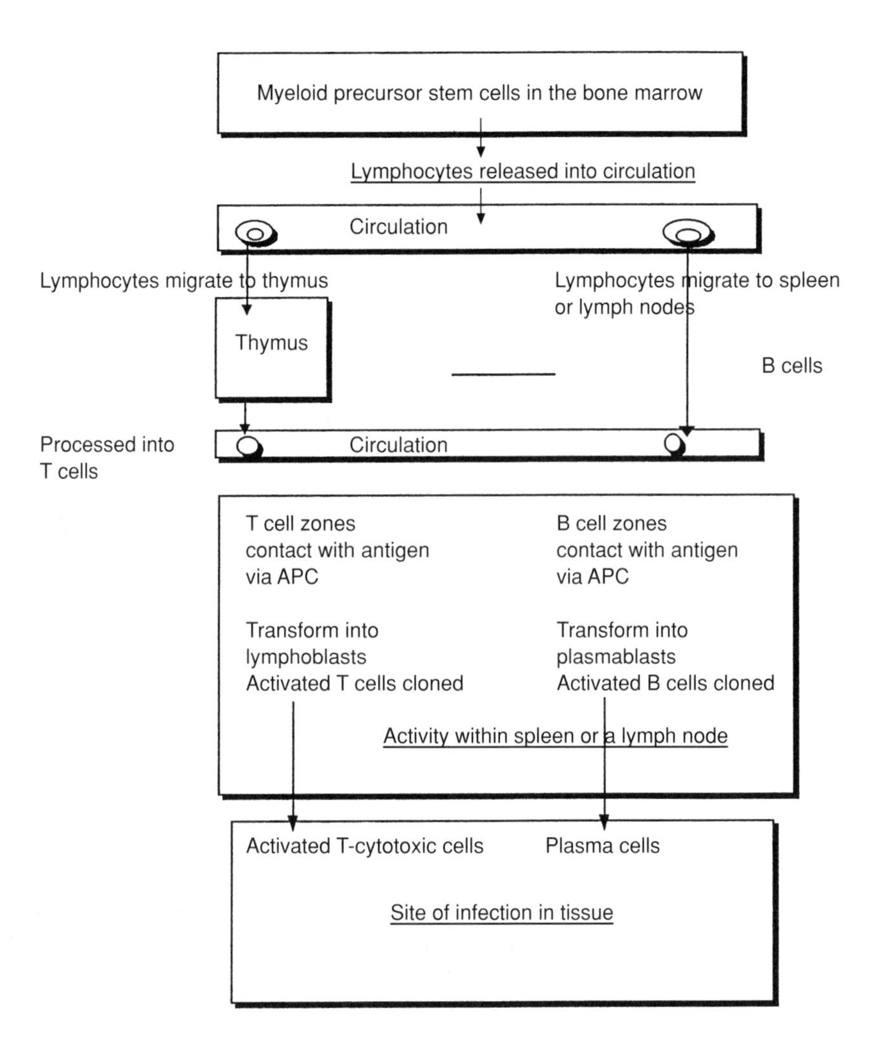

• **Figure 5.5** All the white blood cells (leukocytes) including the lymphocytes originate from bone marrow stem cells. Once the lymphocytes are released into the circulation they are processed into T or B cells. If they proceed directly to the thymus they become T cells. If the lymphocytes migrate first to the spleen or lymph nodes they become B cells.

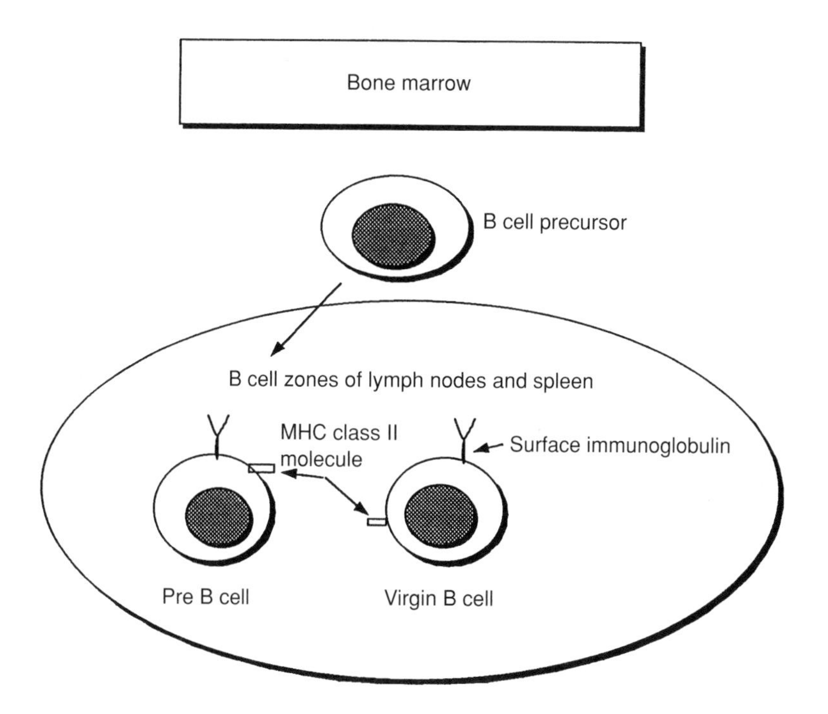

• **Figure 5.6** A precursor B cell matures within the B cell zones in a lymph node or spleen first into a pre B cell and then into a virgin B cell. On the surface of the B cell are membrane/surface immunoglobulin (Ig) molecules and MHC class II molecules. The Ig molecules act as antigen receptors.

The progenitor cells differentiate into lymphoblasts (and these develop into B and T lymphocytes), megakaryocytes (which give rise to platelets/megakaryocytes), monoblasts (which develop into monocytes), and myeloblasts (which give rise to granulocytes).

5.5.1.1 The thymus

The thymus is a primary lymphoid organ where the virgin lymphocytes are processed into mature T lymphocytes (T cells) (antigen independent maturation). It is a bi-lobed organ located between the heart and the sternum, composed of a reticular network of fibrous connective tissue cells surrounded by a capsular membrane. Each lobe is divided into lobules and each lobule has an outer cortex and inner medulla comprised of a reticulum filled with lymphocytes and some epithelial cells.

The virgin lymphocytes congregate in the medulla. Epithelial cells within the thymus produce polypetide hormones that may help the lymphocytes to differentiate (antigen independent differentiation). The lymphocytes accumulate in the cortex and once activated the cells divide by mitosis producing more T lymphocytes (T cells). Only about 5% of T cells exit from the thymus as viable cells to become antigen-reactive cells.

Maturation of T lymphocytes (T cells) begins in the subcapsular region. Cells from the bone marrow, under the influence of epithelial 'nurse cells,' develop into lymphoblasts. Those lymphoblasts in continuous contact with non-lymphoid cells migrate through the cortex into the medulla and mature into T cells (see Fig. 5.7).

5.5.1.2 The spleen

The spleen is the largest secondary lymphoid organ and is located below the stomach and pancreas in the upper left-hand-side of the abdominal cavity. It is a flat dark red organ which can swell to twice its size (splenomegaly) during the course of an infection. The outermost layer consists of a membrane — the outer capsular membrane which surrounds a reticulum of fibroblasts.

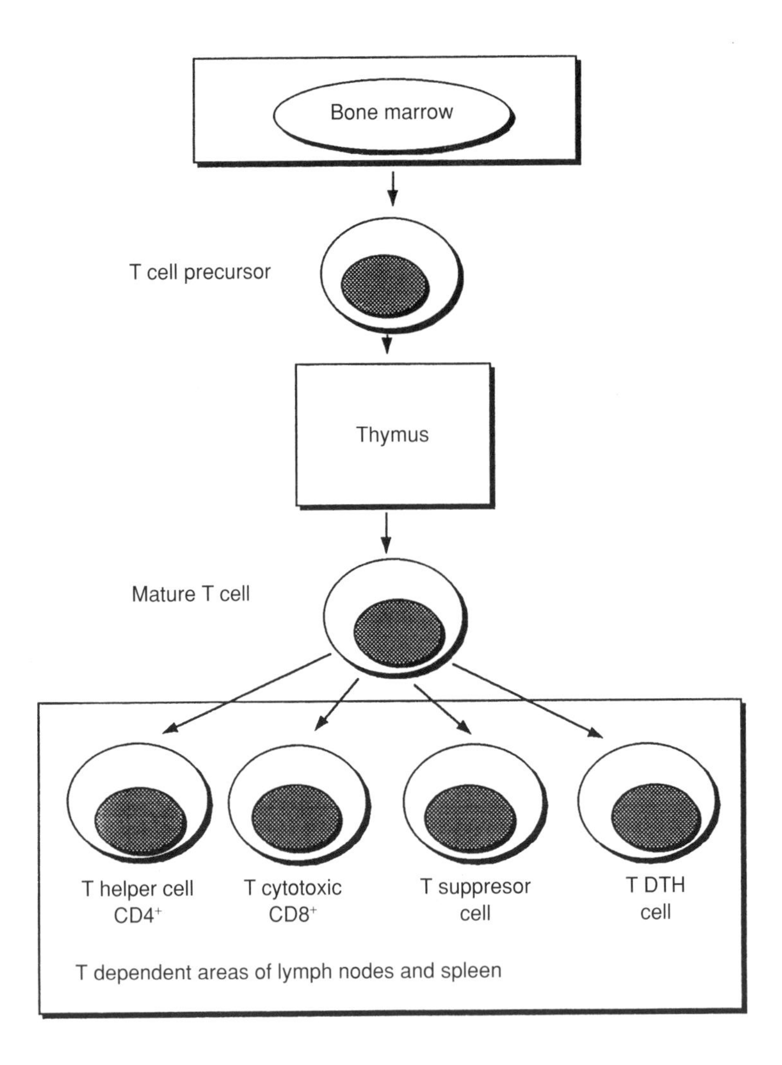

• **Figure 5.7** A precursor T cell migrates to the thymus where it is processed into a mature T cell. Within the T cell zones of lymph nodes and the spleen the mature T cell differentiates into at least four different phenotypes. The T cells have T cell antigen receptors and MHC class I molecules on their surface membranes and T helper cells have MHC class II.

The sinusoids (spaces) are penetrated by arterioles and venules with the capsular membrane penetrating into the reticulum forming trabeculae. There are two distinct regions within the inner zones: the white pulp and the red pulp.

- White pulp is composed mainly of lymphocytes located around arterioles. Immediately surrounding the arteriole is a sheath of T cells known as the periarteriolar sheath (PAS). On the periphery of the PAS are clusters of B cells alongside a germinal centre which produces more lymphocytes. About 50% of the total lymphocytes within the spleen are B cells and 30–40% are T cells.
- Red pulp is composed of a mainly of red blood cells with lymphocytes macrophages and giant cells.

5.5.1.3 The lymph nodes

The lymph nodes are small (about 1 cm diameter) ovoid bodies scattered throughout the body. Each one is supplied with efferent and afferent lymphatic ducts, arteries and veins. An outer capsule encloses a cellular reticulum, interspersed with dendritic lymphocytes and ordinary macrophages.

There are three main cellular zones within each lymph node:

- The outer cortex, a dense cellular area, and within the cortex is the paracortex.
- The innermost layer is the medulla, a central area consisting mostly of a sinus containing the larger vessels.
- Trabeculae — extension of the outer capsule — extend into the node.

The cortex contains germinal centres, where the activated B lymphocytes (B cells) multiply, and follicles (primary lymphoid follicles are composed mainly of B cells). B cells transform into either B memory cells or plasma cells which migrate toward the medullary region. The paracortex is mainly a T cell region. Antigen transported to the spleen by the antigen-presenting cells remains within the paracortex and cortex.

5.6 THE MAIN CELL TYPES INVOLVED WITH THE IMMUNE RESPONSE

Leukocytes are the nucleated cells found in the body fluids, including blood, are classed as either granulocytes or agranulocytes depending upon whether or not they have granules in their cytoplasm that become visible under the light microscope after staining.

- The granulocytes, neutrophils, eosinophils and basophils all arise from the same type of stem cells.
- The agranulocytes are monocytes and lymphocytes and are mainly derived from myeloid stem cells (see Fig. 5.8).

5.6.1 THE GRANULOCYTES

5.6.1.1 Neutrophils

Neutrophils (size about 14 μm), also known as polymorphonuclear leukocytes, have multilobed nuclei with up to five lobes, linked by a chromatin strand. When stained with Romanovsky stains the cytoplasmic granules bind to both acidic and basic dyes and cancel out the staining effect. They then do not show up — hence the name neutrophils. There are two types of granules — primary and secondary — depending upon the time they appear during cell development; and they contain lyosozyme, hydrolases, myeloperoxidases and collegenases.

Neutrophils make up approximately 60–80% of all circulating leukocytes and can penetrate into tissues in response to an infection (the early inflammatory stages). The primary function of neutrophils is phagocytic, that is they digest and destroy pathogenic material, dead host cells and immune complexes by both phagocytosis and pinocytosis and are stimulated by secretions of pathogens, and one of the components of complement. They have Fc receptors binding sites located on their cell membranes which are activated during infections. Neutrophils can release reactive oxygen intermediates (O_2^- radicals).

5.6.1.2 Eosinophils

Eosinophils (size about 16 μm) have a bilobed nucleus and acidophilic granules with a high affinity for the dye eosin (stains pink). Within the cytoplasm are major basic proteins located in ribosomes, mitochondria and granules that contain basic protein substances that, when released, are toxic to pathogens. They constitute 3–5% of the total number of circulating white blood cells. Their numbers tend to increase during allergic reactions, parasitic infections and asthma attacks. The majority of eosinophils are tissue-dwelling cells.

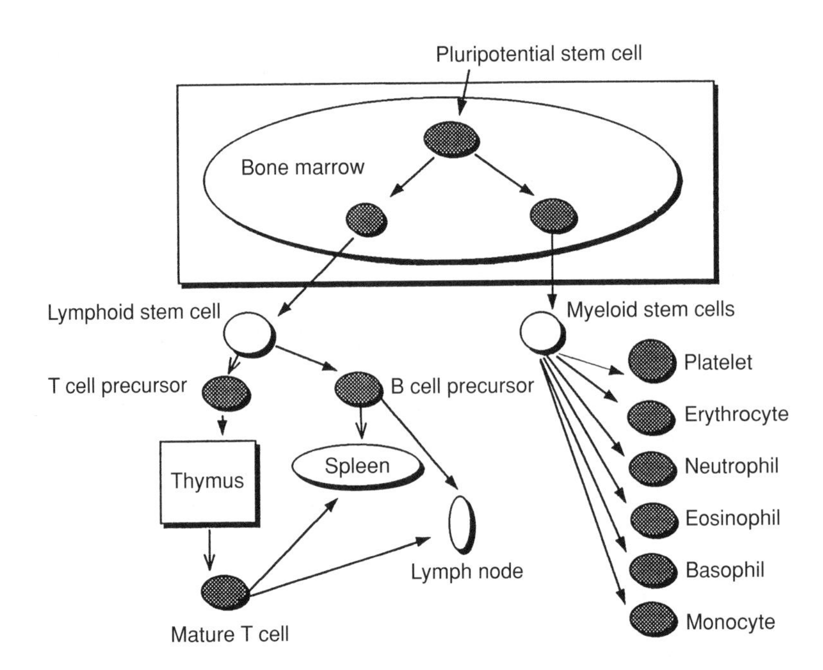

• **Figure 5.8** A basic (pluripotential) stem cell differentiates within the bone marrow into either a myeloid or a lymphoid stem cell. All the lymphocyte phenotypes arise from the lymphoid stem cells and all the other leukocytes originate from the myeloid stem cells.

Eosinophils release activators of hypersensitivity and are phagocytic and can engulf immune complexes and attack weakened parasites. T cells activated by parasitic helminth secretions secrete cytokines some of which stimulate eosinophils to attach to weakened parasites. The eosinophils then discharge their granules onto the parasite tegument.

The antibodies IgG, IgE and IgA attach to activated eosinophil Fc receptors with the Fab end of the antibody molecules attaching to the pathogen. The antibody forms a 'bridge' between the cell and the pathogen and this leads to antibody-dependent cell-mediated cytotoxicity (ADCC). Eosinophils react to histamine and other pharmacologically active substances.

What then is the function of eosinophilia? The results of various experiments described below may provide some insight as to the role of eosinophils in parasitic infections.

Angiostrongylus canonensis is a nematode for which the mouse is a permissive host but the rat is not. In rats the parasites migrate and eventually die. In rats there is a pronounced eosinophilia once the nematode begins to weaken. The parasite survives in mice and the increase in the eosinophil population is far less or almost negligible.

In humans suffering with onchocerciasis during the early stages of the infection, there is no noticeable eosinophilia while the adult worms are still migrating. Once the adult worms nodulate, which might be a sign of ageing, there is an increase in the local eosinophil population.

There is a pentastomid that lives in rattlesnakes and rats. In the rat the parasite undergoes one or two moults. Eosinophils have been observed digesting the dead casts.

All these results suggest that either eosinophils are active against ageing or damaged or dying parasites but not against a 'healthy' living parasite, or a healthy living parasite somehow inhibits the eosinophils.

5.6.1.3 Basophils

Basophils (size 10–14 μm) are granulocytes with a round or ring-shaped nucleus and cytoplasm containing large basophilic granules that stain blue with basic (aniline) dyes. Basophils make up less than 1% of the normal leukocyte component of circulating blood.

The granules contain the precursors for histamine, heparin and other pharmacologically active compounds (they are very similar in function to tissue mast cells). Basophils can be linked to allergens such as pollen grains by an antibody bridge which stimulates them to degranulate.

5.6.2 THE AGRANULOCYTES

5.6.2.1 Monocytes

Monocytes (size 15–30 μm) have a characteristic 'kidney'-shaped nucleus with clear cytoplasm and make up at least 6% of the circulating leukocytes. They are the largest of the white blood cells and can survive for 1–2 months. Monocytes migrate out of the blood vessels into the tissue to become tissue histiocytes (usually at a site of infection). They become either fixed or motile phagocytic cells acting as scavenger cells or antigen-presenting cells (APCs). Monocytes have receptors on the surface of their cell membranes for IgG_{2a}, IgG_{2b} (in mice), IgG and IgE (in humans), cytokines and the C3b component of complement. Activated monocytes produce complement components, interferons, interleukins (mainly IL-1) and prostoglandins.

5.6.2.2 Lymphocytes

Lymphocytes (size 8–12 μm) are the smallest of the white blood cells with a large nucleus to cytoplasm ratio. The have clear cytoplasm with no granules visible under a light microscope after staining. Lymphocytes comprise about 20–30% of the circulating leukocytes and accumulate within the thymus, spleen and lymph nodes. They circulate within the blood vessels and lymphatic ducts and migrate out into tissues when there is an infection.

Lymphocytes originate from haemopoietic stem cells located in the bone marrow. Once the lymphocytes have been differentiated from stem cells in the bone marrow, they migrate to the thymus, lymph nodes and spleen where they divide by mitosis when stimulated. Lymphocytes develop into several different phenotypes depending upon what organ and where in the organ they settle. Each phenotype has a specific function in an immune response.

Those lymphocytes that migrate from the bone marrow directly to the thymus and are 'processed' there, form the 'seed' T lymphocytes (T cells). Lymphocytes that migrate directly to the lymph nodes or spleen from the bone marrow develop into the B lymphocytes (B cells). Within the T cell phenotype there are further subsets — the T helper (Th) cells, T cytotoxic (Tc) cells and T suppressor (Ts) cells.

- B cells after stimulation develop into either plasma (antibody-secreting) cells or B memory cells.
- Both B and T cells have antigen receptors on their surface membranes.
- The B cell antigen receptors are surface (or membrane) immunoglobulins (sIg).
- The T cell the antigen receptor is a molecular complex on the surface membrane known as a CD3 molecule.

Both types of receptor have a 'variable' end that binds to the epitope of the antigen. The structure of the variable end molecule of each individual lymphocyte is unique and hence there are numerous variations.

There is nearly always a circulating lymphocyte that can bind to an antigen that has become associated with a major histocompatibility complex (MHC) class I or class II molecule.

There are different T cell phenotypes which are distinguished from one another by their surface marker molecules. Firstly there are T helper (Th) cells that have a surface molecule known as the CD4 ($CD4^+$) in close association with the CD3 molecule. Secondly there are T cytotoxic (Tc) cells that have CD8 molecules ($CD8^+$) in association with the CD3 receptor and thirdly T suppressor (Ts) cells have both CD4 and CD8 markers.

In addition there is another separate group of lymphocyte type cells: the natural killer (NK) cells, which are granular lymphocyte-like cells.

5.6.3 MACROPHAGES AND PHAGOCYTOSIS

The first cells to respond positively and even aggressively to non-self material are the phagocytic cells. The most common phagocytic cell is a macrophage and originates from a precursor in the bone marrow. From the bone marrow the cell enters the blood circulation as a monocyte and when recruited into the tissues it differentiates into a macrophage.

Phagocytosis is the process whereby the macrophages engulf non-self material including dying and discarded host cells. Macrophages specialise in phagocytosis and have been observed in vitro to engulf microscopic particles as well as microbes. The engulfed microbes are attacked in macrophage cytoplasm by reactive oxygen and nitrogen metabolites.

When macrophages make contact with foreign material they secrete cytokines and chemokines, which are the signals that set an immune response in motion. The first response is that of an innate immune response and if the infection is not resolved it becomes chronic and the response merges into an adaptive immune response. Once the presence of a pathogen is detected both the macrophages become activated and other cells such as mast cells are also stimulated and consequently release chemical mediators of inflammation. Hence the macrophages not only secrete inflammatory mediators but also respond to those derived from other cell sources.

The wound healing that follows inflammation is inaugurated by proteases and growth factors secreted by macrophages. Macrophages engulf, digest and process the pathogen and break it down into peptide fragments. Some of the peptides are transported to the macrophage cell membrane where they become combined with MHC class II molecules. These macrophages migrate to the lymph nodes to become antigen-presenting cells. Those microbes able to survive destruction within the macrophages, that is they can evade the first stage of the immune response, can cause tissue damage and often are the cause of diseases.

5.6.4 MAST CELLS

Mast cells are granular cells that originate from stem cells in the bone marrow. Mature mast cells are nucleated granular cells that congregate in connective tissues throughout the body. These cells play an important role in the both innate and adaptive immune responses. On their external membranes are receptors for antibodies and eosinophils as well as sensors for the detection of any alterations in the form of damage, temperature, oxygen concentration or the presence of pathogens. Any of these changes can stimulate the mast cells to degranulate. The granules contain the precursors to numerous pharmacologically active compounds. Among the most prominent are: histamine, serotonin, bradykinin, 5-hydroxytryptamine (5-HT), slow reacting substance (SRS) etc.

Mastocytosis, an increase in the number of mast cells, is a phenomenon associated with parasitic infections and allergies. As with eosinophils there is a debate about some of the exact functions of mast cells.

It has been demonstrated that there are two distinct mast cell phenotypes based on their biochemical properties:

- Mucosal mast cells which are normally only found in the gut mucosa and contain mast cell protease II.
- Connective tissue mast cells which are present in most tissues of the body outside of the alimentary canal and contain mast cell protease I.

This separation has now been shown to be not quite so simple, both types of mast cells having been shown to be present in the livers of rats infected with tapeworm metacestodes and in connective tissue hyperplasia caused by the infection.

Mast cells have often been linked with the expulsion of gut parasites but the results of numerous investigations into this phenomenon have still not yet demonstrated convincingly whether or not mast cells are the prime factors in the expulsion process.

A case has been made for mast cells damaging both the worms and gut tissue in mice infected with *H. diminuta*.

One of the properties of mast cells is the secretion of angiogenic factors. This results in an increase in the growth of capillaries and blood flow to the infected area. Histamine and other compounds derived from mast cells affect the permeability of the walls of endothelial cells. This allows for a leakage of fluid into the infected zone and the accumulation of muco-polysaccharide compounds, eg glycosaminoglycans (GAG). This creates a semi-viscous environment for the parasite to inhabit. All the nutrients that the parasite requires are probably in solution within the GAG. Hence it may just be that the parasite has become adapted to the host's response and actually uses it to its own advantage.

5.7 THE SECOND PHASE OF THE IMMUNE RESPONSE

There are only a few diseases caused by the types of parasite already outlined that result in the death of the host. That is there are very few parasites that can be considered to be 'killers' like the pathological bacteria and viruses. The phagocytic cells are activated by the presence of a parasite and if they engulf part of or all the parasite they become antigen-presenting cells (APCs).

The presence of activated APCs and cytokines stimulates the T lymphoyctes and in particular the T helper cells. In order for an adaptive immune response to begin both the T cells and B cells have to react to an antigen peptide associated with MHC class II molecules. Both T cells and B cells have antigen receptors on their cell membranes: T cell antigen receptor (TCR) and surface immunoglobulin (sIg) respectively.

Activation of mature T cells leads to the expression of surface molecules. Only those T cells which express receptors that react to non-self antigen in the presence of an MHC molecule are selected. Immature T cells that do not express receptors reacting to self-antigen are eliminated (by clonal selection). The interaction between an antigen peptide associated with an MHC molecule on antigen-presenting cells (APCs) requires additional signals (co-stimulatory signals) provided by co-stimulatory molecules.

The co-stimulatory molecules are thought to have evolved from PAMPs (pathogen associated molecular patterns) on the pathogen and PRRs (pathogen receptor regions) on the host phagocytic cell (see section 5.1). The primary recognition of the pathogen is non-specific for cells already present that have not been specifically cloned to recognise the pathogen. The contact between co-stimulatory molecules stimulates the production of the cytokine IL-2 which then initiates the beginning of an adaptive immune response.

From the above description, it can then be deduced that the main feature of adaptive immunity involves the recognition of an antigen on a specific cell. This response is initiated by the pathogen associated molecular patterns (PAMPs) on the pathogen membrane reacting with pathogen receptors (PRRs) on the phagocytic cells which will become antigen-presenting cells.

The most important result of all this activity is that the macrophages are activated and attempt to destroy the invading parasites, and selected B cells are cloned, producing memory B cells and plasma cells. These then secrete specific antibodies to attach to the invading parasite.

■ 5.7.1 LYMPHOCYTES AND CYTOKINES

Naive lymphocytes (lymhoyctes that have not been exposed to an antigen) are pluripotent; that is they can differentiate along different pathways and develop into effector cells. The effector cells have PRRs on their surface membranes and when stimulated secrete specific cytokines. Cytokines reacting with the T helper cells have a major influence in determining the direction of the adaptive immune response.

■ 5.8 PARASITES, CYTOKINES AND T HELPER CELLS

Relatively recent experimental studies on the immune response to parasites have emphasised the importance of the role of the different T helper cell ($CD4^+$) phenotypes (Th_0, Th_1 and Th_2 cells). A T helper precursor cell (ThP) gives rise to a Th_0 cell which is able to secrete cytokines IL-2, IL-4 and IFN-α. Studies with strains of mice in which Th_1 cells dominate, when infected with the protozoan intracellular parasite *Leishmania major*, have shown that the cytokines they produce are TNF-β, IFN-γ and IL-2, and control the growth of the parasite. Mice with only Th_2 cells, which predominantly secrete IL-4, IL-5, IL-10 and IL-13, experience unrestrained growth of the parasite. In *Leishmania* infections the ability to induce Th_1 and Th_2 cell responses respectively is apparently a balance between IL-12 and IL-4 production.

Parasitic helminths tend to initiate a type II (Th_2) response. In mice infected with *Schistosoma mansoni* experimental evidence indicates that the size of egg granulomas in the liver is modulated by IL-4. The expulsion of the nematode *Heligmosomoides polygyrus* from mice has been shown to be controlled by Th_2 cytokines. The majority of parasitic helminths are extracellular, which apparently does not induce IL-12 production and seems to favour Th_2 cell production.

From the experimental evidence with both protozoan and helminth parasites it appears that the Th_2 cell functions are of primary importance in both immunopathologic and immunoprotective responses to helminth egg granuloma and worm expulsion from the gut.

The protozoan parasites *Toxoplasma* and *Leishmania amastigotes* induce significant IL-12 production when they infect macrophages (mfs). However when *Leishmania* promastigotes infect macrophages they do not induce IL-12 secretion. Promastigotes are killed by activated macrophages but the initial invasion allows time for conversion of the promastigotes to amastigotes without excessive activation of Th_1 cell responses.

On the surface membranes of *Leishmania* promastigotes are molecules of lipophosphoglycan (LPG) which are shed during the invasion of macrophages and their conversion to amastigotes. There is evidence that evasion of IL-12 induction by promastigotes may be due to the presence of the LPG molecules during parasite invasion.

In *S. mansoni* infections there are stage-specific differences in cytokine induction. For example the larvae tend to induce a Th_1 cell response, whilst egg deposition tends to induce a Th$_2$-cell response. In mice infected with *S. mansoni* the egg granuloma formation is a T cell-dependent hypersensitivity reaction mediated by T helper cells (CD4$^+$ Th cells) which are specifically primed for schistosome egg antigen. As the infection proceeds there is a gradual reduction in the size of the granuloma formed around continuously incoming eggs. This may be due to the macrophages failing to stimulate the Th$_1$ cells which subsequently become unresponsive to antigen-presenting cells. This process has been described as immunomodulation and may reflect a state of unresponsiveness of a Th$_1$ subpopulation of CD4$^+$ T cells which are soluble extract of antigen-specific. Hence during the course of a *S. mansoni* infection there appears to be both Th$_1$ and Th$_2$ cell activity. The Th$_1$ cells play a major role in launching the formation of egg granulomas in the early stages of infection at a time when the granuloma size is maximal. Not only do the Th$_1$ cells mediate granuloma size and formation but they also enhance the granulomagenic potential of Th$_2$ cells, also enhancing antibody production and eosinophilia. During the acute phase, however, Th$_1$ cells become unresponsive to stimulation by a new population of antigen-presenting cells. The result is a decreased cytokine production and cell activation, leading to diminished granuloma size (immunomodulation). The antigen-presenting cells (granulate macrophages or accessory cells) are thought to preferentially downregulate Th$_1$ cells, leaving the Th$_2$ cells to sustain antibody production, eosinophilia and reduced granuloma formation.

The T helper cells require two signals in order to carry out their functions. Firstly a primary signal from the complex of MHC class II molecules and antigen-derived peptide to induce mitotic activity, producing a clone of a specific group of T helper cells. Secondly co-stimulatory signals on the antigen-presenting cells are equally necessary for T cell activation.

■ SUMMARY

All vertebrates have the ability to try and protect themselves against invasion by foreign and non-self material, a response referred to as the immune response. There are two basic types of response: the innate and the adaptive immune response. Innate immunity is non-specific and is dependent upon the inherent capability of having phagocytic cells that can recognise non-self molecular patterns on the membranes of invasive pathogens. If the pathogen is not destroyed then the infection becomes chronic and an adaptive immune response develops. This involves the cloning of lymphocytes that have been in contact with the antigen peptides via an antigen-presenting cell. Those lymphocytes are cloned and retain a memory of the antigen's molecular configuration. They subsequently produce specific antibodies and cytotoxic cells to try and control the infection.

The white blood cells and other phagocytic cells are all participants. The cells originate in the bone marrow and are subsequently processed in the thymus. Within the spleen and lymph nodes the lymphocytes make contact with the antigen-presenting cells and the adaptive immune response develops. T helper cells, mast cells and cytokines all play an essential role in the adaptive immune response.

END OF CHAPTER QUESTIONS

INTRODUCTION TO HOST RESPONSE

Question 5.1 In multicellular organisms, how does the genotype of each individual cell help with identification of self?

Question 5.2 What are the innate mechanisms that protect the host from infections?

Question 5.3 What are the differences between an innate and an adaptive immune system?

Question 5.4 What are the possible origins of the vertebrate immune system?

Question 5.5 What local changes occur once a pathogen has penetrated into the dermal layers of the skin?

Question 5.6 Discuss the role of the cell adhesion molecules in inflammation.

Question 5.7 What are mast cells and what are their origins and functions?

Question 5.8 Describe the events that lead up to inflammation.

Question 5.9 What are inflammatory cytokines?

Question 5.10 Which cell types are involved in, and what function do they play, in inflammation?

THE IMMUNE SYSTEM

Question 5.1 Which are the main primary and secondary organs that participate in the immune response?

Question 5.2 What cells are found in the bone marrow and what are their functions?

Question 5.3 What is the fate of a monocyte once it migrates from the blood capillary into the tissues?

Question 5.4 What is the major histocompatibility complex (MHC)?

Question 5.5 How does a digested fragment of pathogen (the antigen peptides) activate a T cell?

Question 5.6 Describe the structure and function of: (a) the thymus; and (b) a typical lymph node.

Question 5.7 Where is the spleen located and what are the main structural features of the spleen?

Question 5.8 What are the functions of the spleen?

Question 5.9 Name the main types of leukocytes, their distinguishing characteristics and their functions.

Question 5.10 Which of the leukocytes has a dominant role in an adaptive immune response?

Question 5.11 What are mast cells and what is their role in an immune response?

Question 5.12 What is an antigen receptor and on which cells are they found?

Question 5.13 Explain how an adaptive immune response is initiated.

Question 5.14 What are costimulatory molecules and what are their origins and functions?

Question 5.15 What is the relationship between PAMPs, PRRs and co-stimulatory molecules?

Question 5.16 Describe cytokines, with examples.

PARASITES AND T HELPER CELLS

Question 5.1 How many different T helper cell phenotypes have been identified?

Question 5.2 What cytokines are secreted by the T helper cells?

Question 5.3 What is the role of the T helper cells in mice infected with *Leishmania major*?

Question 5.4 Which T helper cell phenotype is normally dominant during helminth infections?

Question 5.5 What T cells are involved in the formation of a *Schistosoma mansoni* egg granuloma?

Question 5.6 Outline the activities of the various cytokines in relation to parasite infections.

NUTRITION AND BIOCHEMISTRY OF PARASITES

6

6.1 THE PARASITIC ADVANTAGE

Parasites have the unique advantage over free-living animals in that they live entirely surrounded by their food. The methods of feeding have been adapted to the parasites' habitat. In most cases the food has already been digested into a soluble form available for absorption. Hence some of the helminth groups have dispensed with a gut altogether (the cestodes) and absorb their nutrients through the outer covering (the tegument). Others have a gut but also absorb food through the tegument (the trematodes), while there are some (the nematodes) that only absorb food via their mouths.

The direct uptake of soluble molecules through either a tegument or a membrane is a physiological process similar to that of the uptake of solutes by mammalian gut epithelial cells. The process of absorption centres round the following mechanisms:

- Simple diffusion: molecules are absorbed passively through the membrane: a process regulated by the movement of molecules from a higher concentration into the parasite's cells or tissues with a lower concentration.
- Active transport: absorption whereby molecules pass into the parasite against a concentration gradient. This process requires energy and can be inhibited by substances that interfere with, or inhibit, respiration. This system may involve a 'carrier' molecule and the simultaneous movement of sodium ions.
- Facilitated or mediated diffusion: movement into the parasite by molecules first 'conjugated' to a carrier molecule and then absorbed through the membrane. The solute is released once it has crossed the membrane. The carrier is a locus on the membrane to which the solute binds to and is then released on the other side of the membrane where the concentration is lower.
- Pinocytosis (endocytosis and exocytosis): large molecules such as proteins are transported into the tissues within membrane-bound vesicles.
- Contact digestion: there is evidence to show that absorption of nutrients involves enzymes. These are either intrinsic, ie of parasite origin, or extrinsic, ie of host origin. The intrinsic enzyme phosphohydrolase helps phosphate esters to be absorbed through the tegument. Fructose 1,6-diphosphate is hydrolysed at the tegument surface to release inorganic phosphate that is absorbed while the detached fructose is unable to pass through the tegument.

BOX 6.1 FEEDING

- The cestodes' only means of feeding is by absorption through the tegument.
- Trematodes have a gut and a tegument but it still is a matter for discussion and experimental observation as to how much food is absorbed directly through the tegument and how much via the gut epithelium.
- Nematodes have an outer non-permeable cuticle, an alimentary canal with a 'mouth' and a well developed pharynx to assist in the uptake of food and its distribution through the gut system. The mouthparts of various species are adapted to the different types of feeding, ie whether they feed on tissue or body fluids etc.
- Protozoa fall into two basic categories: those which engulf the food by phagocytosis or pinocytosis; and those which absorb their food through the outer membrane.

6.2 UPTAKE OF NUTRIENTS

A number of studies on the kinetics of the uptake of nutrients have been carried out using two species of adult tapeworm, namely *Hymenolepis diminuta* and *H. microstoma*. Both of these worms can be removed from the gut of their host, a rat and mouse respectively, and maintained in vitro. The cysticerci of *Taenia crassiceps* have also been used to study the uptake of nutrients.

The results obtained from those experiments indicate that the uptake of nutrients by those parasites follow similar pathways to the manner in which nutrients are absorbed into the gut epithelial cells. Parameters such as pH, temperature, oxygen, carbon dioxide and enzymes can all be influential in the absorption of soluble nutrients. The methods used for the study of rates of enzyme reaction such as Lineweaver–Burke plots, from which maximum uptake and affinity rates can be calculated, have been applied to the study of a parasite's nutrient uptake.

- Transport kinetics reveal that the uptake of carbohydrates is dependent on the molecular shape and structure of the carbohydrate molecule, and often there is competition for the identical transport locus (competitive inhibition).
- Glucose absorbed by active transport in the presence of galactose or fructose may lead to competitive inhibition of the uptake of compounds, particularly if one of them is at a higher concentration.

6.2.1 UPTAKE OF AMINO-ACIDS

Amino-acids, from the gut amino-acid pool, are absorbed by tapeworms. There appears to be neutral, basic and acidic amino-acid transport systems. The amino-acids enter the worms via mediated transport systems and accumulate against a concentration gradient. Cestodes unlike mammals require sodium and chloride ions coupled to the amino-acid transport system. The cestodes have an equal affinity for the absorption of both D and L amino-acids. However apparently a two-way process is in operation. Molecules leak from as well as enter the parasite. A pool of amino-acids is usually present in the lumen of the host's gut and a similar 'pool' exists within the tissues of the parasite. The movement of amino-acids in and out of the parasite maintains an equilibrium between the amino-acid pool in the host's gut lumen and the pool within the parasite.

Cestodes are not able to synthesise cholesterol. In order to overcome this deficiency solutions of mono-olein and sodium taurocholate are absorbed. The tapeworm *H. diminuta* has an uptake site for a short-chain (C_2–C_8) and a separate one for long-chain (C_{14}–C_{24}) fatty acids. This system can only function in the presence of the host's bile salts. *H. diminuta* cannot synthesise long-chain fatty acids but can lengthen the chain of absorbed fatty acids by adding acetate units.

Parasites that are closely related phylogenetically do not necessarily have similar feeding mechanisms. The absorption and uptake of nutrients are often determined by the nature of the habitat, for example *Fasciola hepatica* and *Schistosoma mansoni* are both trematodes but one lives in the bile duct and the other in mesenteric blood vessels. However, *F. hepatica* absorbs amino-acids by passive diffusion whereas *S. mansoni* apparently uses an active transport system to absorb amino-acids.

Feeding mechanisms are often adapted to their immediate environment as illustrated by the nematode *Ancylostoma duodenale* (a hookworm). Like all hookworms *A. duodenale* is a gut-dwelling nematode that is equipped with biting mouth parts and apparently can secrete an anti-clotting agent to prevent the host's blood from clotting before it can be swallowed.

6.2.2 ENERGY STORAGE

All parasites need a supply of energy to maintain and regulate their metabolism. It has been established that many parasitic helminths can function for limited periods under anaerobic conditions. All require a source of reduced organic compounds and a mechanism for the release and capture of energy (see Fig. 6.1). In most helminths the energy is stored as glycogen and, in the protozoans *Plasmodium* and *Trypanosoma*, carbohydrate is the main energy store whereas in *Entamoeba* it is probably glycogen.

The parasites investigated from a biochemical point of view indicate that they conform to most of the established biochemical pathways. All of the parasites observed oxidise glucose by the same glycolytic pathway as that encountered in free-living organisms. The oxidation pathway that produces phosphoenolpyruvate (PEP) is common to nearly all parasites but the conversion of PEP to pyruvate and further degradations are not common. Some follow the mammalian patterns, other do not. In mammals PEP is converted to pyruvate and then normally passes to the tricarboxylic acid cycle. This cycle is far less active in parasites and when it does operate, it does so at a very low level and appears to be far less important to parasites than to mammals. One of the reasons for this is that parasites do not need to expend much energy in their search for food.

6.2.2.1 Energy for reproduction

The helminths have, in general, complex sets of reproductive organs that produce numerous eggs with protective sets of membranes and shells. In many of the nematodes, the females give birth to live larvae. Adult cestodes, trematodes and nematodes have a high fecundity rate and some of the metacestodes reproduce asexually. Most energy production, whether it be from an aerobic or anaerobic source, is required to fuel the reproductive process. In addition there are added demands such as the production of hatching enzymes and penetrating enzymes.

- Many of the passive stages such as cysts, metacercaria and eggs require the appropriate external stimulus before they develop any further.
- Schistosome eggs need a temperature lower than of the host, together with both light and water in order to hatch.

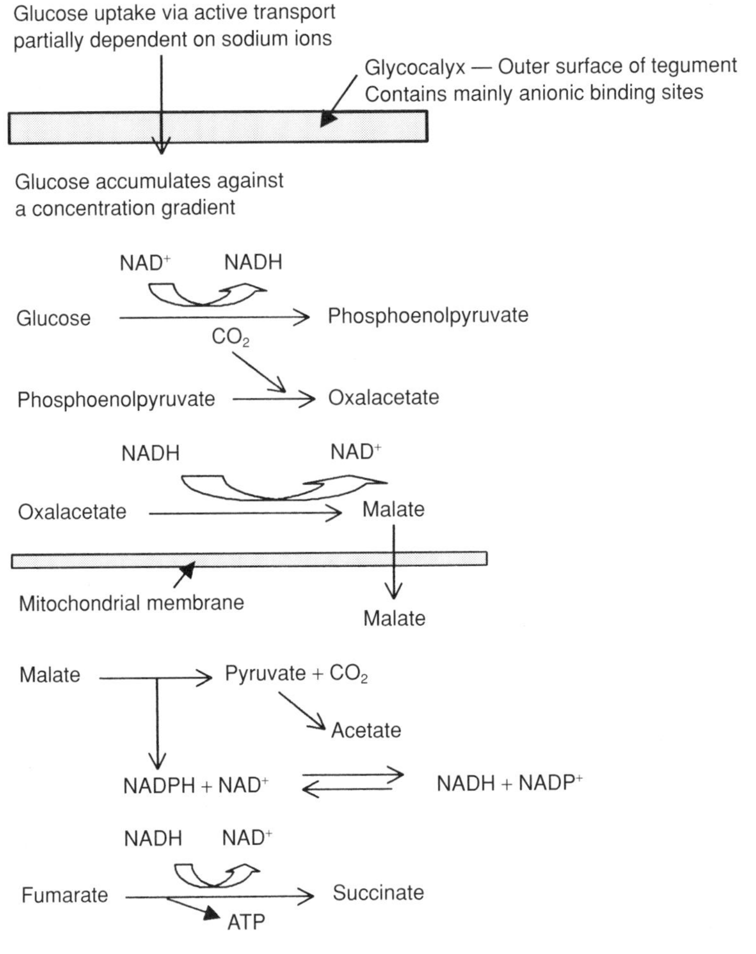

• **Figure 6.1** The capture, storage and active transport of energy in parasitic helminths.

In parasitic playthelminths glucose is absorbed from an exogenous source mainly by a process of active transport. Having passed through the tegument the glucose is stored mainly as glycogen within the parenchyma cells.
When glucose is required the glycogen is converted into glucose.
The glucose is then converted into malate and lactate via phosphoenolpyruvate and oxalacetate.
The malate is absorbed through the mitochondrion membrane and broken down into acetate and lactate via phosphoenolpyruvate and oxalacetate. Malate is broken down into acetate and succinate.
The later reaction produces the energy-rich compound ATP.

NAD = nicotinamide adenine trinucleotide; NADH = the reduced form of NAD; ATP = adenosine triphosphate — an energy-rich compound.

Adapted from Carbohydrate Metabolism by L.S. Roberts in: Biology of the Eucestoda; edited by C. Arme & P.W. Papas (1983). Published by Academic Press.

- *Eimeria* cysts only release their sporozoites after they have been subjected to the conditions of first the mammalian stomach and then the small intestine.
- Cestode eggs only hatch once inside the appropriate host and the larvae normally require the gut's digestive enzymes, particularly bile, to develop into the next stage.

6.3 BASIC PHYSIOLOGY AND METABOLIC PATHWAYS

Each individual parasite has evolved and adapted basic physiological and metabolic pathways for its own needs. There are numerous reports and reviews which deal in detail with the variations and similarities of different parasites. Some of the more important features of parasites in relation to their metabolism and physiology are:

- The cestodes and trematodes are triplobastic without a body cavity. They are aceolomate and all movement of solutes is by diffusion from cell to cell or through the mesodermal matrix.
- The nematodes have a body cavity, the pseudocoelome, which is filled with pseudocoelomic fluid (also known as perienteric fluid haemolymph). The osmotic pressure of the fluid is equivalent to 0.2M NaCl, the pH is between 6.2–6.4; and within the fluid proteins, fats, carbohydrates, enzymes, organic acids and small quantities of haemoglobin are stored.
- The role of haemoglobin in nematodes is not clear and the haemoglobin does not deoxygenate at low pressures. In *Ascaris* the body haemaglobin deoxygenates slowly under aerobic conditions.
- Among the carbohydrates are glucose (trace quantities), the disaccharide trehalose and glycogen. Glycogen is the main polysaccharide and is stored in the body wall and muscle tissues.
- Keratin, sclerotin and collagen are the main structural proteins. The collagen exists in two forms: (1) basement membrane collagen; and (2) cuticular collagens.

All classes of neutral lipid and phospholipid have been identified in different nematode species. In eggs and in the reproductive tract of female *Ascaris* a unique series of α-glycosides have been isolated.

The protein, lipid and carbohydrate composition in cestodes is different from most invertebrates. The glycogen (carbohydrate) level is relatively high and the protein level relatively low. Investigations have shown that metacestodes contain the highest levels of glycogen.

The body structure of both trematodes and cestodes contains calcareous corpuscles ranging in size over 12–32 μm. These structures are made from both organic base and inorganic material. The corpuscles originate intracellularly — one per cell which is eventually destroyed. The role of these structures is the subject of much speculation. One view is that they act as a buffer for acids produced by anaerobic respiration and another is that they act as reserves of phosphates, other organic ions and carbon dioxide.

6.4 TRANSMISSION OF PARASITES BETWEEN HOSTS

The manner in which parasites are distributed from one host to another is often reflected in their physiological systems. This topic has been dealt with in more detail in Chapters 1 and 4.

6.4.1 ESTABLISHMENT

The survival of a parasite depends upon locating a suitable host and then overcoming the host's defences. Once inside the host, the correct physiological conditions such as pH, O_2 and temperature must be stimulated for hatching/exsheathing. In the cestodes bile salts also play a very important role in the following ways:

BOX 6.2 A SUMMARY OF THE BASIC TRANSMISSION SYSTEMS

- The egg or infective larval stage is eaten by the potential definitive host.
- The egg or larval stage is eaten by the intermediate host. The larval stage that develops within the intermediate host is shed and eaten by the definitive host. The intermediate host is the prey of the definitive host.
- Free-living larval stages shed by the intermediate host then actively penetrate the definitive host.
- Transmission from one host to another via vectors.
- Transmission can be affected by circadian rhythms. A circadian rhythm is based on a 24 h cycle and certain parasites have become adapted to this system:
  - Rhythms associated with synchronous cell division, eg the malaria parasites.
  - Rhythms associated with the discharge of infective forms either from the definitive host, eg *Coccidia*, *Enterobius* and schistosomes or from an intermediate host, eg Schistosomes.
  - Rhythms associated with migration of the parasite, eg Trypanosomes, malaria parasites and nematode microfilaria.
  - No rhythms of any sort have been observed among the cestodes.

- Effects on membrane permeability.
- Initiation of activity of the larva/metacestode.
- Lytic effect on parasite surfaces.
- Synergistic action with host digestive enzymes.
- Effects on the metabolism of the establishing parasite.

A bile salt molecule is composed of a hydrophilic and a hydrophobic domain and reacts with both the lipids and proteins in the membrane thereby affecting the permeability. This allows water and enzymes to assist the hatching of eggs or excystation of cysts.

The deoxycholic acid in bile salts may influence the host specificity of certain cestodes. Deoxycholic acid has been shown to be able to lyse protoscoleces. Rabbits and sheep are rich in deoxycholic acid and hence the protoscoleces of *E. granulosus* do not survive in those hosts. Dogs, a permissive host for *E. granulosus*, have much less deoxchyolate acid in their bile salts.

6.4.2 FACTORS AFFECTING DEVELOPMENT

Most parasites require a signal from the host to begin their development. A clear example of this is in the case of hydatid cysts of *E. granulosus*. If in the intermediate host the cyst is damaged in any way and protoscoleces escape, but remain within the intermediate host, they just form secondary cysts.

The numbers of worms per host or the biomass of worms can influence the effect of the parasite upon the host but also can affect the parasite. In laboratory infections of rats with *Hymenolepis diminuta* it has been demonstrated that the mean mass of worms decreases as the number of worms per host increases. Relatively recently it has been shown that the greater the number of adults present in the gut the greater is the effect upon the length and morphology of proglottids (Stradowski 1996).

■ **BOX 6.3**

Cornford (1991) studied the glucose transporter kinetics in *H. diminuta* and observed the following results after culturing the worms in vitro.

The Michaelis constant (K_m) for 8-day-old worms was K_m = 0.34 mM, and the maximum uptake rate(V_{max}) was V_{max} = 14 nmol per min per g

For 10-day old worms the K_m = 0.46 mM, and the V_{max} = 18 nmol per min per g

For 17-day old worms the K_m = 0.51 mM, and the V_{max} = 21 nmol per min per g

The age of the worms appears to be a factor affecting uptake and it could also apply to drug uptake. An older parasite may be more susceptible than younger parasites and the dose rate could be altered accordingly, especially where the drug has side effects upon the host.

■ **BOX 6.4**

Ito et al. (1988) studied antibody responses in *H. nana*. In susceptible mice they found antibodies to eggs, cysticercoids and adults. In a BALB/c non-susceptible strain of mice (that is, they expel the adult worms after 30 days) antibodies were produced to egg antigens only. However worms were rejected when there were few or no detectable antibodies in the serum. This suggests that antibodies alone do not have an effective role in the expulsion of parasites and that other aspects of the immune system may be involved. The different antibody responses suggest that there is a difference in antigen specificity between eggs, cysts and adults. This latter observation may have to be taken into account in attempts to develop vaccines and prophylactic drugs.

The in vivo and in vitro culture used in studying the kinetics of nutrient uptake can also be applied to drug studies particularly the uptake and the biochemical effects of the drug.

McCracken and Taylor (1983) examined what effect in vivo, then in vitro, the drugs (benzimidazole anthelmintics) thiabendazole (TBZ) and cambendazole (CBZ) had upon *H. diminuta*. Their results demonstrated that a single oral dose of TBZ, at a dose rate of 250 mg/kg, on day 15 eliminated 100% of the worms.

Some the observed effects upon the worms were:

- Changes in weight and chemical composition.
- They were significantly smaller and with a smaller percentage body weight glycogen.
- The protein concentrations had increased to offset the decline in glycogen.
- The glycogen/protein ratio in TBZ treated worms was lower than in control worms.
- CBZ was found to be five times more potent than TBZ against *H. diminuta* and produced the same basic changes within 18 h of treatment.
- In vitro studies of treated worms showed that they metabolised smaller quantities of exogenous glucose than controls.
- The ability to accumulate glucose against a concentration difference was significantly depressed.

■ BOX 6.5

Hildreth et al. (1997) showed that the drug tunicamycin at a dose of 10 mg/ml significantly inhibited the uptake of tritiated galactose into the worm.

Tunicamycin inhibits the synthesis of glycoproteins, procollagen and peptidoglycans and the secretion of IgA and IgE by plasma cells. In addition the compound also affects the fibroblast interferon cell system. The level of inhibitory effect of TM appears to be at the level of protein glycosylation rather than carbohydrate (galactose) transport.

McCracken and Lipkowitz (1990) investigated the effect of tioxidazole (TIOX) on *H. diminuta* infected rats. The following observations were made:

- There were alterations to the weight of the worms as well as their chemical composition. The worms were found to be significantly smaller and contained less glycogen.
- The protein concentrations increased to offset the drop in glycogen.
- Glycogen/protein ratio was less than in controls.
- There were differences in absolute amounts of protein and glycogen compared to controls.

During in vitro studies using TIOX treated worms, the worms absorbed less exogenous glucose and the ability to absorb against a concentration gradient was significantly depressed.

■ 6.5 ANTIPARASITIC CHEMOTHERAPY

The structure of the drug molecules influences the affect they have on tapeworms. For example the molecular configurations of benzothiazole and benzimidazole are structurally related and both are active in causing irreversible damage to the tegument of tapeworms.

Many parasites, especially the malaria parasites, have been shown to have developed resistance to some of the more commonly used drugs. Among the parasitic helminths the mechanism of resistance includes parasite-specific enzymes that are able to protect against oxygen radicals and develop resistance against drugs acting via an oxidative burst.

In many of the helminths, especially the nematodes, the microtubules and nervous system appear to be the main chemotherapeutic targets in helminths. They differ from those of the host because of the evolutionary distance separating mammals from helminths. However one of the more recent discoveries has been that the microtubules of free-living nematodes, eg *Caenorhabditis elegans* are just as susceptible to benzimidazole anthelmintics as parasitic nematodes. The motoneural map for *C. elegans* is the same as that for *Ascaris lumbricoides*. Both are immobilised by levamisole, piperazine, avermectins etc. This interesting finding could possibly lead to many more free-living nematodes being used for testing new drugs.

Problems with the drugs used against parasitic protozoa give a high potential for mutation, which many of these parasites undergo. A rapidly forming new generation of parasites provides the conditions for the possibility of drug resistance to develop. The parasite causing cerebral malaria, *Plasmodium falciparum*, is known to have developed resistance to some of the anti-malarials now available.

Phosphofructokinase (PFK) extracted from schistosomes is sensitive to antimonial compounds which may due to the PFK from parasites being an isoenzyme of mammalian PFK.

■ BOX 6.6

Drug	Effective against
Chloroquine phosphate	*Plamodium* spp (Malaria therapy & prophylaxis)
Diethylacarbamazine	Filarial nematodes
Ivermectin	Filarial nematodes
Mebendazole	Most cestodes
Albendazole	Mainly cestodes
Praziquantel	Mainly *Schistosoma* spp

■ SUMMARY

Parasites live surrounded by their food and hence they have reduced the musculature and sense required in searching for food. The gut-dwelling parasites are immersed in digested and semi-digested food. The nutrients are generally absorbed as amino-acid, carbohydrate or fat molecules either by passive or active absorption. The kinetics of absorption are similar to those of the cells that line the gut of the host.

A great deal of the energy requirement of the parasite relates to maintaining its large fecundity — the ability to produce large numbers of eggs or larvae. The metabolic pathways that are part of protein synthesis and energy synthesis are similar to those of most vertebrates. In many cases parasites have become facultative anaerobes. The chemical signals and stimuli that are required for establishing growth and development are obtained from the host.

Knowledge of differences between the parasite's and the host's biochemical pathways can be explored for the possible development of drugs.

END OF CHAPTER QUESTIONS

Question 6.1 What is the main difference with regard to feeding between a free-living animal and a parasitic animal?

Question 6.2 What are the main feeding methods found in the protozoa, cestodes, trematodes and nematodes?

Question 6.3 What is the difference between passive uptake and active uptake?

Question 6.4 Describe mediated or facilitated uptake, contact digestion and pinocytosis.

Question 6.5 What factors affect the uptake of nutrients by cestodes?

Question 6.6 Describe the uptake of glucose by cestodes.

Question 6.7 How are amino-acids absorbed by tapeworms?

Question 6.8 How do trematodes absorb their nutrients?

Question 6.9 What feeding mechanisms are used by nematodes?

Question 6.10 Outline the methods of respiration that occur in the protozoa, cestodes, trematodes and nematodes.

Question 6.11 Describe what functions of a parasitic helminth use most of the metabolic energy produced.

Question 6.12 Describe some of the more important features of parasites in relation to the metabolism and physiology.

Question 6.13 How does the oxidation of glucose pathway observed in helminths differ from that seen in vertebrates?

Question 6.14 Outline the various ways in which the basic metabolic pathways have been adapted to the parasites' requirements.

Question 6.15 What are calcareous corpuscles and where are they found?

Question 6.16 Describe the various modes of transmission from one host to another.

Question 6.17 What are the factors that assist in the establishment of parasites?

Question 6.18 What factors can affect the development of parasites?

Question 6.19 How has the uptake of nutrients by cestodes been investigated?

Question 6.20 Describe some of the effects that anti-helminth drugs can have upon the parasites.

PATHOLOGICAL EFFECT OF THE PARASITE UPON THE HOST

7

■ 7.1 INVADING THE HOST'S BODY

The parasites' habitat, whether it is inter- or intracellular, will be affected by the presence of the parasite. Simple movement through an organ or tissues is disruptive and damaging, which has to be corrected. Once a parasite has invaded a cell both the metabolism and the function of that cell are disrupted. The parasite absorbs its nutrients from the cytoplasm and then deposits its metabolic waste into the cytoplasm. For example the *Plasmodium* species that infects humans feeds on haemaglobin during the erythrocytic phase. The partially digested pigment is deposited in the liver and spleen as haematin and the organs take on a characteristic brown coloration. The nitrogenous waste produced by the parasite's metabolism is simply passed into the host and, if present in sufficient quantities, will become toxic to the host.

The majority of intracellular protozoan parasites, after having invaded a host cell (an erythrocyte or a gut epithelial cell), undergo a multiplicative phase (eg *Plasmodium*, *Eimeria*, *Leishmania*). The progeny of the invasive stage escape from the cell to invade other similar cells. The host cell is irreversibly damaged during this process and, as the parasitaemia increases, more and more cells are invaded and consequently an increasing number are destroyed. This leads to either the necrosis of surrounding tissue or the leaking of body fluids, particularly if the epthithelial cells of vessels are damaged, or anaemia if blood cells are destroyed.

Gut-dwelling nematodes damage the lining of the lumen either by becoming embedded or by their feeding activities. Frequently the epithelial cells are damaged resulting in alterations to the gut epithelium architecture and function. Often such changes to the gut mucosa are the cause of malabsorption syndromes. Accumulation of numerous large nematodes in the gut apart from causing malfunctioning, can also block the lumen, a condition that can be fatal.

Individual cells destroyed by intracellular parasites are eventually replaced in a normal healthy host and normally there is no lasting effect. However when tissue cells are damaged either by feeding, parasite embedment or migrations, those cells are not always replaced by identical cells but by fibrous or scar tissue. The immediate host reaction to the presence of the parasites (see section 5.3) is the accumulation of phagocytic cells and the release of chemical signals that initiate the onset of an inflammatory response.

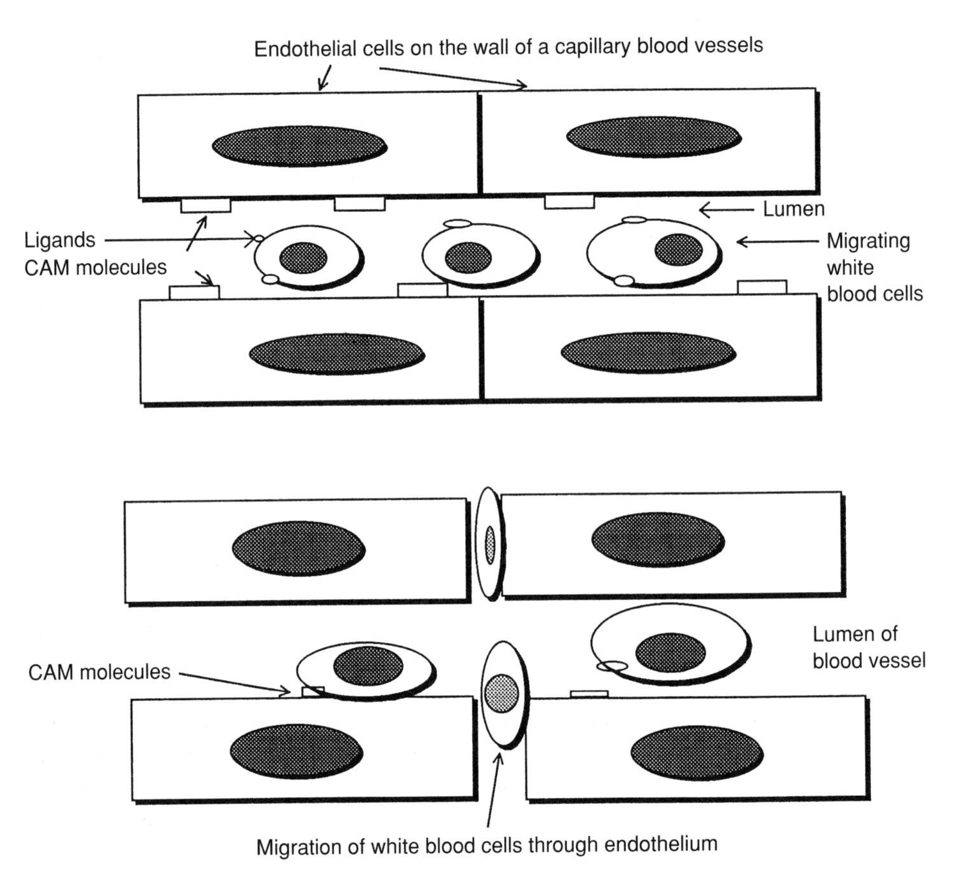

• Figure 7.1 The leukocytes circulate round the body in the blood vessels. The endothelial cells on the walls of blood vessels have various cell adhesion molecules (CAM) on their membranes. On the surface of the leukocytes are CAM receptor molecules known as ligands.

• Figure 7.2 During an infectious state the CAM molecules on the endothelial cells of the nearest blood vessels are activated. The ligands on the leukocytes become receptive and the leukocytes then accumulate by adhering to the CAM molecules. The walls of the endothelial cells are affected by various vaso-active factors released from locally activated mast cells and this then permits the leukocytes to migrate from the blood vessels into the tissues to the site of the infection.

■ 7.2 INFLAMMATION CAUSED BY PARASITES

In tissues damaged by physical injury or pathogen invasion the tissue mast cells degranulate, releasing histamine and serotonin, vasoactive amines. Those compounds cause relaxation of the smooth muscles of the capillaries and arterioles and affect the permeability of membranes of local cells. The vasoactive amines loosen the adhesions between adjacent cell membranes, forming gaps between adjacent epithelial cells. Blood plasma now leaks into the surrounding tissue and leukocytes migrate from the lumen of the vessels into the tissues (a process known as diapedesis, see Fig. 7.1). There is an increase in blood flow to the infected site or injury. This is followed by a local temperature increase due to the secretions of pyrogens secreted by leukocytes (mainly neutrophils) and redness due to leakage of body fluids.

The leakage of fluids and cells is drawn by chemotactic and cytokine stimuli to the site of parasite invasion. The local cells are activated by the presence of the pathogen and secrete various chemotactic stimulants to attract leukocytes into the area. Circulating leukocytes are attracted to the site of infection by the pathogen-activated cell adhesion molecules (CAM) on the membranes of epithelial cells. On the external cell membranes of circulating leukocytes are ligands — molecules which when activated bind to the appropriate cell adhesion molecules (see section 5.7). Tissue damage and movement through the tissue disrupts local nerve endings. The subsequent neurological response is a contraction of local smooth blood vessels and a temporary alteration in the blood flow.

The macroscopic signs of inflammation at the site of parasite invasion are erythema (redness), swelling, pain, heat: all related to alterations in microcirculation and the change in permeability of the walls of epithelial cells.

Vasoactive hormones such as prostaglandins (PGE) and leukotrienes (LT) are released by the mast cells into the surrounding tissues. These are lipid hormones derived from arachidonic acid, a compound that originates by the action of phospholipase on the membrane phospholipids. Any substance that damages cell membranes activates the phospholipase.

- The major prostaglandins are PGE1, PGE2, $PGE2_a$, and thromboxane A2. The prostaglandins were first isolated from seminal fluid produced by the prostrate gland and they are produced by most cells including phagocytic white blood cells.
- The leukotrienes — include LTB4, LTC4, LTE4 and LTF4 and are derived from white blood cells (leukocytes).

The pain associated with inflammation:

- is due partly to swelling, the influx of fluids causing pressure on the local nerve endings;
- and also to direct action from vasoactive amines such as histamine, serotonin and bradykinin.

The precursors to those compounds are located in the mast cell granules (see section 5.6.4) and are released when the mast cells are stimulated to degranulate. These amines act together with (synergistically) with prostaglandins and this induces pain. Pain acts as a signal to the system that there is damage and that the area needs some protecting.

In addition to what has just been described, more than one blood protein system is activated by injury or infection. Once a precursor of these blood proteins is activated they set off a series of chemical reactions known as a chemical cascade and the reactions generate soluble mediators of inflammation. The purpose of the cascade is to amplify the local response and to initiate the subsequent events that occur away from the site.

Specific plasma proteins involved in the chemical cascades associated with inflammation include kinins, the complement systems, the clotting and the fibrinolytic systems. The first of the blood plasma proteins to be activated is known as the Hageman Factor (HFa) which is the first component factor (Factor XII) of the clotting system. Activated HFa stimulates the kinin and fibrinolytic systems.

Collagen, a chemical compound commonly found in all higher animals, forms a matrix in which all body cells are embedded and when tissue is damaged the Hageman factor is activated and binds to the collagen in the damaged tissue. For the healing or resolving process to proceed the resultant clot must be dissolved. This is achieved by the enzymatic activities of plasmin. As the fibrin network is slowly degraded the phagocytic cells attack any trapped microbes. As the clot disappears it is replaced by scar tissue. The scar tissue is formed by fibre type cells and deposits of elastin and collagen.

About 10% of the globulin fraction of blood serum is made up of a multicomponent system known as complement, consisting of enzymes and binding proteins. Once activated the complement cascade generates mediators of inflammation and some of these are distinct and others are identical to other systems involved. Complement can be activated directly by a microbe — the alternate pathway — or via the immune system — the classical pathway. The end from either pathway is the accumulation of cells associated with inflammation and phagocytosis.

During the inflammatory response the local activated macrophages release an endogenous pyrogen which acts on the part of the brain — the hypothalamus — that controls the body temperature. The result is a slight increase in body temperature — a fever, a symptom of inflammation. During inflammation the stimulated macrophages secrete the cytokine interleukin-1 (IL-1) which has been shown to induce fever and stimulate lymphocytes (T cells). Another cell secretion released into the blood serum during inflammation is colony-stimulating factor (CSF) which induces the production of leukocytes.

During the initial stages of inflammation the system generates products that firstly enhance inflammation and then the same products serve as feedback to control the inflammatory process. An example are the acute-phase reactants, a group of plasma proteins synthesised by liver cells (hepatocytes) within 6–12 h of the commencement of the immune response. These proteins serve mainly as protease inhibitors that try to limit the damage to tissues.

■ 7.3 REACTION TO PARASITIC HELMINTHS

Once a helminth, whether an adult, juvenile or larval stage, enters the body proper (ie it does not remain within the lumen of the alimentary canal), a host response is activated. However most parasites have adapted to this situation (Fig. 1.2) and have evolved different methods of avoiding the host's immediate response. As long as the parasite remains healthy and virile, the protective devices seem to be able to protect the parasite.

Host phagocytic cells try to adhere to the parasite but generally without any real impact. If the parasite is damaged in any way, or becomes sluggish through age, many of these host cells adhere and attempt to invade the parasite through its cuticle or tegument.

Neutrophils and tissue monocytes (histocytes) are usually the first cells on the scene and, if they are activated, the parasite becomes sluggish and the neutrophils are replaced by eosinophils. These cells and any other phagocytic cells that may become attached release compounds that contain either reactive nitrogen and/or reactive oxygen intermediates. These molecules damage the parasites and if their concentration increases they can even become destructive to host tissue. Eosinophils adhere to the surface of parasites and degranulate, releasing ingredients such as major basic protein (MBP), which is destructive to parasite tissue. At this stage the parasite may be weakened by age or simply not properly adapted to the environment.

The number of adhering cells increases with new cells replacing the spent ones. These cells gradually destroy the internal cells of the parasite and its tissues begin to degenerate. As destruction of the parasite increases the eosinophils on the outside are replaced by epitheloid and fibrocyte (fibre-forming) cells.

With nematodes the cuticle often remains intact although the internal tissues degenerate, whereas in the tapeworm metacestode and the trematode adult and juvenile stages the tegument also disintegrates. Eventually a fibrous capsule forms a round the dead worm. Once the capsulation process is complete calcium is deposited in the capsule layers.

The system just described can be observed under experimental conditions. The filarial nematode *Dipetelonema viteae* can survive subcutaneously in a hamster — a permissive host, but not in a mouse — a non-permissive host. If live adult worms are surgically transplanted just beneath the dermal layers of a mouse the process of inflammation and gradual encapsulation of the parasite begins and, at about 20 days post-infection, the worms are dead and completely encapsulated. If worms are killed by immersion in formaldehyde before transplantation, then the process of inflammation and encapsulation of the worm begins after 1–2 days.

A similar scenario can be observed when the protoscoleces of *Echinoccocus granulosus* are transplanted directly or within a diffusion chamber into the peritoneal cavity of a mouse. The end product is a very definite macroscopic capsule surrounding a protoscolex. However a small percentage of live protoscoleces can survive when the mouse is infected with a relatively large number of juvenile tapeworms.

The eggs of *Schistosoma mansoni* that travel via the blood circulation and end up in the hepatic tissue remain there. A granuloma (see section 7.5) develops round the egg and eventually eosinophils can be observed in close contact with the egg, which is then destroyed.

However with some helminths the host can react to the presence of a live healthy parasite and encapsulate it, but the parasite within the host capsule remains alive.

The tetrathyridia of *Mesocestoides corti* can live free in the peritoneal cavity and can also invade the liver tissue. The hepatic cells along the migratory path are necrosed and replaced with cells associated with inflammation. Eventually the tract is filled with collagen and fibrin (scar tissue). Once the tetrathyridium has settled, a host capsule forms round it. This type of capsule differs from the granuloma that forms round schistosome eggs. The parasite remains active within a semi-fluid mucilaginous medium. The inner boundary of the capsule is composed of regular epithelial cells and outside of these are layers of fibrocytes among deposits of collagen. Outside of this layer are found eosinophils, lymphocytes, macrophages and mast cells. No host cells can normally be seen attached to the parasite.

If the same host is infected with both *M. corti* and *S. mansoni* both tetrathryridia and eggs can be seen within the liver. However the eggs are surrounded by a typical granuloma whereas the tetrathyridia stay alive within a host capsule. This is evidence for the fact that parasites appear to have a great deal of influence in controlling the nature of the host response.

If the eggs of *Taenia solium* are accidentally swallowed by man (the wrong host) they develop into cysticerci and migrate round the body. Often they settle in organs such as the brain. Once settled, the host reacts with an innate inflammatory response. A fibrous capsule eventually forms round the cyst which may increase in size with time. The enlarging 'cyst' causes the surrounding brain tissue to become necrotic and impairs aspects of the functioning of the central nervous system. This condition is known as neurocysticercosis and one of the symptoms of this infection is very similar to epilepsy.

Epileptic fits due to neurocysticercosis is a not uncommon condition encountered in areas such as Mexico, India, vast regions of Africa and S. America where *T. solium* is endemic.

■ 7.4 ORGAN AND SYSTEMIC PATHOLOGY

The spleen is a lymphoid organ that reacts uniformly to all types of infection. Within the white pulp of the spleen there is a great deal of activity related to an adaptive immune response. An increase in the amount of white pulp area is probably due to the increased numbers of lymphocytes (both T and B cells) cloned in response to the infection. The giant cells found in the red pulp are probably a fusion of individual macrophages and also increase in number. This may be due to the fact that the infection is producing excess waste products that need to be detoxified and removed from the system.

The outer capsular layer of the liver, spleen and visceral organs that comes in contact with the parasites becomes thickened and fibrosed. If there is a heavy infection of loose parasites within the peritoneal cavity, the fibrin deposits on the surface of the organs cause the organs to adhere to one another.

The presence of a parasite within the abdominal cavity stimulates a general inflammatory response with an increase in fluid-secreting ascites cells. Hence there is an accumulation of fluid within the peritoneal cavity, a condition referred to as oedema. Parasitised organs increase in size and often show signs of a colour change or even decolorisation. Often within the organ the pathology is localised arround the parasite.

The effects of protozoan parasites are varied; they could be localised as with *Eimeria tenella*, a parasite of the gut mucosal cells. The resultant pathology disrupts the bowel functions. *Trypanosoma cruzi* in the extracellular phase has a systemic effect upon the host, whereas when it is in the tissue phase it damages local tissues, usually the heart muscle cells. *Plasmodium*, a parasite which is systemic, has an overall effect upon the host rather than a specific site effect. Whether the actual damage is local or systemic the presence of parasites has an overall effect upon the host, usually in the form of fever, nausea, pain, dyspepsia or immunosuppression.

7.5 HUMAN TREMATODE PATHOGENS

7.5.1 *PARAGONIMUS WESTERMANI* (LUNG FLUKE)

The adult *Paragonimus* lives in the human lung and produces eggs after 16 days. Eggs pass out of the body via the sputum. The eggs need to make contact with water before they hatch and, like most trematode eggs, require an optimum temperature of 25–28°C to hatch (always different from that of the host). In water miracidia hatch out of the eggs and swim around until they make contact with a gastropod (Pleuroceridae and Thiaridae only). Once contact is made the miracidia penetrate through the foot and migrate toward the digestive gland. Within the miracidium (see section 3.10) are germ cells which develop into sporocysts (the so-called mother sporocysts) and likewise the sporocysts contain germ cells which develop into the next phase, the rediae. From within the rediae cercariae of the small microcercous type are shed into the surrounding aquatic environment. The cercariae have a small tail, a well developed stylet and oral sucker, and can penetrate into freshwater crab and crayfish muscles where they encyst into metacercariae. Once the crustacean's muscles are eaten the cysts are digested and release juvenile flukes into the intestinal lumen. The juvenile forms burrow through the intestinal wall into the body cavity. They migrate toward the thorax and then through the diaphragm into the pleural cavity and settle in the lungs, where they develop and mature into adult forms.

7.5.2 *OPISTHORCHIS VIVERRINI* (THE CAUSE OF CHOLANGIOCARCINOMA) AND *CLONORCHIS SINENSIS* (THE CAUSE OF CARCINOMA OF THE BILE DUCT)

O. viverrini

The adult parasites live in the human bile duct. Eggs pass out via the faeces but only hatch when eaten by a snail (*Bythnia* spp). A sporocyst (the mother sporocyst) develops within the snail. The sporocyst produces rediae and they develop into cercariae which pass out of the snail. These cercariae have a longitudinal tail fin (gymnocephalous) and swim freely until they make contact with a fish. The tail drops off and they penetrate into the fish muscles and encyst (a metacercaria). The metacercariae are ingested when raw or semi-cooked fish is eaten. The metacercariae excyst in the lumen of the small intestine, releasing juveniles that then migrate up the biliary tree to the small intrahepatic bile ducts, where they mature.

The infection is entirely mucosal and there are no migration phases. Mature worms never actually enter the tissues. Mature worms produce eggs which move down the biliary tract and are excreted with the faeces.

Prevalence in snails is about 1% but nevertheless it was found that in certain areas of South East Asia 100% of certain cyprinoid species harbour metacercariae. Reports of considerable variation in the prevalence and number of metacerariae in fish are associated with the time of year and the species.

Both parasites respond to praziquantel treatment.

O. viverrini (liver fluke) occurs in N.E. Thailand and Loas and infects an estimated 7 million people. *Clonorchis sinensis* and *O. felinus* occur in S.E. Asia, the former Soviet Union and E. Europe. About 20 million people worldwide are infected with these flukes.

The incidence of human infection is dependent on the season, source of species and individual fish consumed. In Thailand there is consumption of raw fish and in N.E. Laos the descendants of Thai people also consume large amounts of raw fish.

Most children become infected by the age of ten. Some adults show intensity of egg output with age that may reflect a continual accumulation of parasites with no immunity. Decline in egg output over the age of 50 could be due to a change in behaviour and usage of praziquantel. Decline in infections with age might also reflect parasite-associated mortality. Studies of infection in hamsters indicate little evidence of immunity to the parasite.

The observed worm burdens were found to be highly overdispersed, like most other helminths in human communities. Using expulsive chemotherapy 11,000 *Opisthorchis* worms were recovered from 246 infected residents of one village. 81% of the worms were expelled by 25 individuals with burdens of over 100 worms. During an examination of 181 fatal accident victims, 66% of worms came from 30 heavily infected cadavers.

■ 7.5.3 CHOLANGIOCARCINOMA (CANCER OF THE GALL BLADDER)

There is some evidence for a relationship between cholangiocarcinoma and the major human liver flukes *O. viverrini* and *C. sinenis*. The mechanism of the onset of cholangiocarcinoma has been investigated using infected hamsters. Possibly immnuocompromised, the infected hamsters develop the disease following exposure to normally sub-carcinogenic doses of *N*-nitroso-compounds or their precursors. However, with humans there may be consumption of low levels of nitrosamines via foods preserved by bacterial fermentation (an important component of the diet) which may be the primary carcinogen leading to cholangiocarcinoma.

In response to *O. viverrini*, the biliary epithelium may become susceptible to malignant transformation due to dietary carcinogens causing chronic proliferation of epithelial cells. *N*-nitroso-compounds produced endogenously by activated macrophages in the chronically inflamed biliary tract may serve as carcinogens. In fluke-infected bile duct the activated macrophages may produce similar carcinogens.

In choliangiocarcinoma once the tumour enlarges patients become wasted and jaundiced and suffer a painful death, usually due to a condition known as ascending cholangitis. Other biliary and gall bladder diseases such as cholangitis, obstructive jaundice, cholelithiasis (stones in the gall bladder) and a cholecystic condition (chronic inflammation of the gall bladder) are associated with liver flukes.

Diagnosis is mainly looking through stool samples for eggs. Each worm produces approximately 50–120 eggs/g per stool. The correlation between worm burden and

■ BOX 7.1

There is a 12-fold higher frequency of this cancer in endemic areas compared to the rest of Thailand. In western countries the incidence of liver cancers is about 2/100,000 population whereas in Khon Kaen Province the occurrence of liver cancers is 89.2/100,000 in males and 35.5/100,000 in females. New studies carried out using ultrasonography and stool examinations show that cholangiocarcinoma may be higher in the general population than as shown in hospital patients.

Praziquantel at a dose of 40 mg/kg body weight is highly effective against most pathogenic trematodes. Treatment also appears to reverse parasite gall bladder abnormalities as observed by ultrasonagraphy.

The disease bilharzia (schistosomiasis) is caused by the trematode parasite *Schistosoma* spp. Urinary schistosomiasis is caused by *S. haematobium* and bowel schistosomiasis by *S. mansoni*, *S. japonicum* and *S. intercalatum*.

antibodies is weak. Parasite DNA is detectable by dot-blot hybridisation using a radio-labelled probe consisting of 340 bases from repetitive sequences of the parasite genome. Parasite antibodies are detectable by the ELISA technique using a mixture of three monoclonal antibodies against a 89 kDa secreted component of the adult worm.

■ 7.5.4 *SCHISTOSOMA* SPP

The life-cycle and basic biology of the schistosomes is described in section 4.8.1. The disease caused by these parasites is known as bilharzia, named after the person who first described this disease. The adult stages of *S. mansoni* can be easily maintained in laboratory mice and primates and the larval stages in the snail *Biomphalaria glabrata*.

An important aspect of the understanding of how this parasite is able to avoid the host's response has come about via the surgical transplanting of adult worms from mice to monkeys. If the worms were transferred from a mouse to a monkey they survived, but if the monkey was sensitised/immunised against mouse proteins the parasites did not survive. This implied that the parasite had somehow or other acquired host antigens/proteins onto its tegument. Hence the monkey's immune system recognised the worms as mouse tissue. The monkey had been immunised to be anti-mouse and hence rejected the worms.

From the above series of experiments it was deduced that during a primary infection it takes the host about 7+ days for the mouse to develop an adaptive immune response involving the production of anti-worm antibodies and sensitised immune cells. During this period the parasite has sufficient time to absorb host proteins onto its tegument which provide the parasite with a molecular disguise. Thus it is not recognised as foreign by the adaptive immune system. The parasites from the primary infection survive but the host is now primed/sensitised and should there be a secondary infection the host is now able to mount a more rapid attack on the parasite and is apparently able to destroy most of the secondary invaders.

This system from the parasite's point of view prevents the host from becoming overcrowded and as a result both the host and the parasite are able to survive. The host is partially protected or immunised by having a live parasite, a condition known as non-sterile immunity. The type of immune response centres round the production of antibodies against adult worm antigens.

However the surviving parasites mature and eventually produce eggs which pass out of the host via the faeces. Some of these eggs migrate to the liver where they stimulate a cell-mediated immune response leading to the formation of a granuloma (see section 7.5). If soluble extracts of the egg (soluble egg antigen) is introduced into the hepatic tissue, a cell-mediated response is invoked as well as the formation of granulomas. This indicates that egg antigen and adult worm antigen are different, each provoking a different type of immune response.

Egg granulomas cause a certain amount of damage and malfunctioning of certain aspects of the liver, for instance interference in bile production and the destruction of hepatic tissue which is replaced with collagen and fibrin to form scar tissue. The pathology of the type just described is a result of the host's immune response and is often referred to as immunopathology.

■ 7.5.5 *SCHISTOSOMA MANSONI*, *S. JAPONICUM* AND *S. INTERCALATUM*

The adults of these three species of *Schistosoma* are found in the portal and mesenteric blood vessels. Once the cercariae have made contact with the skin they begin to penet-

■ BOX 7.2

A post-mortem examination of Zimbabwean women with *S. haematobium* found that urinary schistosomiasis was the cause of the following pathology: genital lesions induced by sequestered eggs and lesions in the vagina, fallopian tubes, ovaries, corpus uteri, and parametrium.

Lesions of the upper genital tract may cause retarded intra-uterine foetal growth and pre-term deliveries. The clinical manifestations a mainly dyspareuria, dysmenorrhoea, bloody vaginal discharge, primary and secondary infertility and ectopic pregnancy. A relationship between genital schistosomaisis and cervical cancer has been suggested (Richter J. et al. 1995).

rate into the dermal layers. This causes a certain amount of itching ('swimmers itch') and may last for 2–3 days.

Once the juvenile (the schistosomule) is established in the dermal tissue it then begins a slow migratory process. The first destination is the lung and symptoms associated with this stage are eosinophilia, fever, diarrhoea and coughing. At this stage the lymph nodes and spleen increase in size. The overall symptom is referred to as Katayama fever and is normally not fatal. The infected individual recovers, but is often left with a chronic infection. Once the worms reach maturity in the mesenteric blood vessels egg production commences.

The egg granulomas that form within the liver lead to portal fibrosis and an increase in the size of the liver and there is portal hypertension. Eggs can be deposited in the lungs and the granulomas can lead to obstructive disease of the lungs. In rare instances eggs are deposited in the spinal cord and brain where the inflammatory response leads to necrosis of a certain amount of neural tissue.

■ 7.5.6 *SCHISTOSOMA HAEMATOBIUM*

S. haematobium adult worms live in the blood vessels associated with the urinary tract and bladder. Unlike *S. mansoni*, *S. japonicum* and *S. intercalatum* the pathology due to this infection (*S. haematobium*) is mainly related to the urino-genital tract.

The eggs pass out of the host via the urine. In order to enter the lumen of the bladder the eggs have to penetrate through the walls (the endothelial cell layer) of the blood vessels. This process causes local haemorrhage and blood loss via the urine. Many of the eggs deposited in the urino-genital tract become embedded in the lining walls and granulomatous tissue forms round the eggs. As the disease progresses there is calcium deposition, detectable as radiopaque patches in the urinary tract.

Continuous 'trickle' infections lead to an inflammation of the bladder and even to the formation of malignant squamous epithelial cells. The inflammatory changes in the ureter followed by fibrosis often lead to obstruction and possible kidney damage. Immune complex deposits have been detected on the kidney glomerular membranes.

■ 7.5.7 *FASCIOLA HEPATICA* AND FASCIOLIASIS

An adult *Fasciola hepatica* has a flat leaf-like shape (Fig. 3.7) with dimensions of about 20–30 × 8–13 mm. *Fasciola* like most of the trematodes is an hermaphrodite. The distribution of the male and female genitalia follows the typical trematode pattern. The outer covering is a tegument and the anterior portion is covered in spines. The anterior end has a cone with an anterior sucker surrounding the mouth. There is a second sucker, the posterior sucker, located a short way behind the anterior sucker. *Fasciola* has a blind-ending gut, hence faeces pass out via the oral aperture.

The excretory system consists of flame cells. Each flame cell encloses cilia which beat and drive fluid from the cell into collecting tubules. The tubules lead into a network of ducts that finally ends in a secretory vesicle.

The tegument is used for both absorption and excretion. The surface of the tegument is continuously shed and replaced. The antigenicity of the fluke changes with the changing tegument.

7.5.7.1 Life-cycle

An adult *Fasciola* lives in the large bile duct and migrates to the smaller bile ducts to feed (see Fig. 4.11). The adult is hermaphrodite, eggs are excreted into the bile and exit the host via the faeces. The eggs are embryonated and take up to 15 days to mature in fresh water. A free-swimming ciliated miracidium hatches out of the egg. When the miracidium makes contact with an aquatic Lymnaeid mollusc, it penetrates the foot of the snail and migrates to the hepatopancreas (digestive gland) where future development takes place. Each miracidium contains germ and somatic cells. The next generation, the sporocysts, develops from the germ cells. Germ cells within the sporocyst form the next larval phase, the redia.

The cercariae develop from germ cells within the redia and escape from the snail via an exit pore. Cercariae at first are free-swimming and then settle onto vegetation and each encysts into a metacercaria which has a life span of 2–3 months. Development from egg to metacercaria takes about 4–7 weeks.

Metacercariae when eaten by the definitive host (an herbivorous mammal) excyst in the duodenum releasing a juvenile stage. The juvenile stage penetrates the intestinal wall and migrates through the peritoneal cavity to the liver where it matures into an adult fluke.

Adult flukes can produce 500–700 eggs per day, with more eggs being produced in the morning, which possibly hints at periodicity. The most common definitive host is sheep. However a range of mammals including man can also serve as the definitive host.

7.5.7.2 Prevalence

BOX 7.3

Fascioliasis is a disease of economic importance. In the 19th century *Fasiciola hepatica* was estimated to be responsible for loss of one million sheep per year. During 1956–57 there was an outbreak of human fascioliasis (about 500 cases) in Europe. In 1969 in Australia sheep farmers suffered large losses due to fascioliasis. The most common cause of human fascioliasis is eating contaminated watercress.

Fascioliasis as an infection has been known for about 600 years, and was first recorded in 1379 in sheep by De Brie.

- 1532 The liver fluke was described by Fitzherbert
- 1698 Fluke eggs were discovered by Bidloo
- 1758 The genus *Fasciola* was created by Linnaeus, who named the fluke *Fasciola hepatica*
- 1818 Rediae were discovered in freshwater snails by Ludwig Bojanus

- 1837 Ciliated larvae from *F. hepatica* were observed by Freidich Creplin
- 1852 Fluke eggs do not cause an infection, as shown by Simonds
- 1875 Cercariae are released from *Lymnaea truncatula* and seen to climb up vegetation, possibly to encyst. This was noted by David Weinland, who speculated that they may be ingested by sheep
- 1882 Development in snail noted by Rudolf Leukart
- 1882 Algernon Thomas observed that miracidia developed into sporocysts within the snail and that rediae are produced from the sporocyst. He also showed that rediae can reproduce and that cercariae with tails are formed in the rediae. He also suggested that cercariae encysted on the grass
- 1914 True migration pathway through the definitive host demonstrated by Dimitry Szinitzin

7.5.7.3 The disease fascioliasis

Two species of *Fasciola* (*F. hepatica* and *F. gigantica*) infect humans. The disease has been reported in 56 countries and highest of the recent outbreaks occurred in Bolivia, Peru, Portugal and Egypt. The damage caused by the infection increases the possibility of liver cancer.

- Symptoms: Loss of weight; accumulation of fluids in the abdomen due to liver damage; 'bottle-jaw' or collection of fluid in the lower jaw; diarrhoea; lethargy followed by death.
- The extent of the disease depends upon the number of challenging metacercariae. Ingestion of 200–800 metacercariae causes chronic disease; 800–1,500 causes sub-acute disease and 800–4,000 causes acute disease. Ingestion of 100 or more adult flukes is usually lethal.

Entry into the liver causes acute traumatic hepatitis. The outer capsule (Glisson's capsule) becomes irregular and fibrosed due to fluke migration. At 10 days post-infection there is an inflammatory response in some sub-capsular regions of the liver close to penetration sites. Migration through the hepatic tissue leaves a trail of necrotic tracts. The migratory tunnels can cause haemorrhagic lesions surrounded by an extensive gathering of neutrophils, eosinophils, lymphocytes, macrophages and hepatocytes. However, the leukocytes never seem to be close to the parasite.

During chronic and secondary infections neutrophils are replace by eosinophils. The increase in weight of the lymph nodes and particular swellings of the follicles suggest that there is an increase in lymphocyte activity, possibly B cells.

Toxic products, possibly metabolic waste, are released by the parasite and could be the possible cause of some of the liver tissue necrosis. This damage is in addition to the mechanical damage caused by the parasite's migratory activities, such as peri-portal fibrosis accompanied by bile duct proliferation. The fibrosis leads to a distortion of the hepatic architecture.

The fluke's entry into the bile ducts produces new symptoms such as hyperplastic cholangitis (proliferation and size increase of the bile ducts) and secretion of large amounts of proline by the parasite, which irritates the ducts. Expansion of the bile duct occurs resulting in an exudation and emigration of cells into the surrounding tissues (lamina propria and surrounding adventitia). In some advanced cases granuloma-like reactions form round the fluke eggs in the smaller bile ducts resulting in the destruction and fibrosis of the ducts. Adult flukes in the bile ducts penetrate the duct walls and feed off blood. Excessive feeding can result in the host suffering from anaemia.

Often in heavily infected sheep secondary infections are fatal. The anaerobic bacterium *Clostridium oedematiens* is often present but dormant in the liver. It is activated as a result of the host's resistance being lowered by the presence of flukes. The inhibitory factors that apparently control bacterial growth are removed and bacteria are able to multiply, producing toxins and resulting in 'black disease'.

In light infections, eggs are found in the faeces after about 56 days.

7.5.7.4 Human fascioliasis

In humans a light infection (ie a relatively small number of flukes) can cause considerable tissue reaction; and calcification in the bile ducts has been observed. Three phases of the disease in humans have been described:

1. Migratory or acute phase when the flukes migrate through the liver characterised by fever, gastrointestinal disturbances and abdominal pain.
2. Latent phase — once the flukes have entered the bile ducts — no pain or symptoms.
3. Obstructive or chronic phase when the adult flukes living in the bile ducts cause swelling of the duct and local liver tissue, with inflammation and obstruction.

Diagnosis: The following methods are used to diagnose fascioliasis: eggs in faeces; complement fixation tests; ELISA; indirect haemagglutination tests; ultrasound.

During the initial and acute phases the symptoms are fever, enlargement of the liver (hepatomegaly), abdominal pain, anorexia, wasting and skin irritation (urticaria). Bleeding bile ducts may lead to anaemia. There are also acute ectopic forms which occur when the juvenile flukes deviate from their usual path of migration and enter organs other than the liver, including the brain.

The chronic form of human fascioliasis is an obstruction of the extrahepatic bile ducts causing biliary colic, jaundice, bouts of cholangitis and pancreatisis.

7.6 THE BIOLOGY OF *ECHINOCOCCUS*

The adults of *E. multilocularis* are found in the alimentary canal of carnivores, mostly foxes although the domestic dog can also be a definitive host. Small rodents such as voles, mice, and squirrels are the natural intermediate hosts. However man can also serve as an intermediate host and human infections have been reported in Europe, Russia and N. America. The eggs, voided via the faeces, are ovoid single oncospheres surrounded by a thick embryophore. The intermediate host becomes infected by ingesting eggs and, when activated by bile, the eggs hatch out in the gut. The hatched oncospheres penetrate the intestinal villi and enter the venous and lymphatic vessels. The majority of oncospheres develop into a coenurus cyst (a metacestode, see section 3.8.4) which migrates to the liver, although other sites such as the brain are not uncommon. A coenurus cyst has a vesicular structure and the oncospheres vesicularises in the liver (or other sites) and develops by asexual proliferation. It takes about 28 days or more for the ceonurus cyst to develop.

The coenurus develops into a fluid-filled vesicle with an inner germinative syncitial layer and an outer laminated layer. The germinative layer produces exogenous buds which form into secondary cysts (referred to as vesicles) and this leads to a mass expansion of the original cyst (vesicular and metastasis formation). Production of protoscoleces from the germinative layer within individual vesicles commences and a multi-vesicular cyst develops within 1–4 months.

■ **BOX 7.4**

The disease caused by *E. multilocularis* shows up in different forms which are referred to as disease states:

- Patients susceptible to the disease develop intensive proliferating intra-hepatic *E. multilocularis* alveolar cysts. Circulating antibodies known as anti-Em-2 (echinoccosis membrane) antibodies can be detected in the body fluids. There is an increase in the number of lymphocytes, particularly cytotoxic T lymphocytes ($CD8^+$ T cells). When removed from the patient these respond to *E. multilocularis* antigen in vitro. In addition, high levels of IL-5 responsible for eosinophil production are found in the serum of such patients.
- Persons who are susceptible to the infection but resistant to the disease develop calcified lesions that do not contain living parasites. The debris from the laminated layer stimulates and maintains the production of antibodies against antigen (Em-2). There is a high proliferative response of lymphoycytes to Em-2 antigen resulting in an increased number of T helper cells ($CD4^+$ T cells).
- Persons who are immunologically resistant to the infection. These are healthy individuals with serological activity to Em-2 antigen but no signs of disease. There is no post-oncospheral development and these individuals appear to be immunologically resistant to infection. But if their serum antibodies are tested, they can respond specifically to the parasite.
- Persons who are constitutionally resistant to the infection. They have a high rate of exposure to the infection but show no immunological reaction.

■ 7.6.1 INFECTION IN HUMANS

In humans (intermediate hosts) *E. multilocularis* is found exclusively in the liver and is referred to as alveolar echinococcosis, or alveolar hydatid disease.

The metacestode progressively invades surrounding tissue by forming a protrusion from the germinal layer, which protects itself from the host response by synthesis of an outer laminated layer. A hepatic lesion forms which macroscopically appears as a dispersed mass of tissue consisting of small cysts and vesicles. The lesions caused by the parasite's development form focal zones of calcification and necrosis. Microscopic examination shows evidence of proliferation of host fibrous tissue, reflecting the host tissue's response to the metacestode.

The clinical signs of the disease are hepatic carcinoma and cirrhosis. Immunodiagnosis is based on the presence of a specific antibody known as the Em-2 antibody, detected by the ELISA technique. The Em-2 antibody is directed against purified *E. multilocularis* carbohydrate antigen. Parasitic lesions can be detected by computed tomography and show up as indistinct and ill-delineated solid masses.

■ 7.6.2 ANTIBODIES

During the course of the infection all of the different classes of immunoglobulins are detected. In particular there is an increased serum level of IgE as well as an increase in the level of IgE bound to circulating basophils. The antibodies may restrict the growth of the metacestode and are probably involved in immunopathological mechanisms, resulting in a certain amount of granulomatous tissue. There are amyloid and immune complex deposits in the infected liver.

■ BOX 7.5

Hydatidosis is the disease caused by this parasite and has been known for a long time. Hydatid cysts recovered form humans were first described by Pallas in 1766. In 1782 protoscolices were first described by Goeze who found them within a cyst. Adult worms recovered from dogs were first described in 1853 by von Siebold and this was followed by the 1863 by Nauyn who fed dogs with the contents of hydatid cysts obtained from humans and recovered adult worms.

■ 7.6.3 BIOLOGY OF *E. GRANULOSUS*

This cestode is associated with domestic animals and humans; and has a worldwide distribution. The adult *E. granulosus* consists of only three proglottids, with a total length of 5 mm; and inhabits the small intestine of the canine family, dogs, jackals and hyenas. The metacestodes, known as hydatid cysts, develop from hatched eggs within intermediate hosts, which are mainly herbivorous animals and man. The most common intermediate hosts are sheep, horses and cattle.

In a mature adult worm eggs accumulate in the posterior segment/proglottid which when ripe (gravid) disintegrates, releasing the eggs which pass out of the host via the faeces. The eggs containing the oncosphere larva are swallowed by grazing animals. The oncosphere hatches out into the small intestine, burrows into the gut wall and migrates round the body via the blood circulation. The liver and lungs are the most common organs in which the larvae settle, however they sometimes find their way into other organs.

Once the larva has reached a suitable site, it settles and secretes a surrounding hyaline membrane. The membrane undergoes a process of development and differentiates into an inner germinal layer and an outer acellular laminate layer. This structure, which has an outer appearance of a small bladder, becomes fluid-filled and over a period of time enlarges into a hydatid cyst. The innermost layer has two functions: it forms the germinal layer which produces the protoscoleces; and secretes more of the outer layer.

As the cyst increases in size the germinal layer loops inwards to form brood capsules and, in each of the brood capsules, the germinal membrane continues to produce protoscoleces. Eventually a large fluid-filled cyst develops containing numerous brood capsules.

It can take several months for a cyst to mature and they vary in size, 2–20 cm. Once a hydatid cyst is eaten by the definitive host, the numerous protoscoleces are released into the host's intestine, become attached to the gut mucosa by means of hooks and suckers and each grows into an adult tapeworm.

Adult tapeworms appear to have little or no obvious effect upon the host. However in the intermediate host the hydatid cyst increases in size and its presence causes the disease known as hydatidosis. Should the cyst rupture or be damaged in any way, the released brood capsules can establish themselves in new tissues and form into secondary cysts.

A growing hydatid cyst is first surrounded by inflammatory tissue and the tissue in which it is embedded often becomes soft, tender and necrotic. Eventually the inflammatory cells are replaced by fibrous/scar tissue and the infected area hardens.

In the lungs, the pulmonary tissue adjacent to the cysts is replaced by inflammatory tissue and thickens into layers of collagen and fibrogen. Histological observations show that the host tissue adjacent to the cyst usually consists of nothing more than layers of fibres. The cellular nature of the tissue becomes more apparent as the histological section moves into the actual lung tissue. Where the hydatid membrane is adjacent to the fibrous tissue, the lamination is more intense and there is some evidence that keratin-like material is deposited.

- *E. granulosus* cysts are unilocular whereas the cysts of *E. multilocularis* are alveolar — formed by series of cysts that have budded off from the original mother cyst.
- An alveolar cyst can cause far more pathological damage to the host, particularly if it becomes established in the brain tissues.

7.6.3.1 Diagnosis

Detecting the presence of a hydatid cyst in the intermediate host and in particular in humans can be done in one of the following ways:

- Infected lungs are detected by radiography and examination of the sputum for protoscoleces or hooklets.
- Ultrasound scanning of the liver (now one of the most widely used methods of diagnosis).
- Immunodiagnosis using a variation of the ELISA (enzyme linked immunosorbent assay) technique. There is evidence to indicate that in older infections the concentrations of IgG_4 antibodies increase with the age of infection.

7.6.3.2 Strains of *E. granulosus*

Chemical analysis has been carried out on *E. granulosus* obtained from different geographical locations. Differences in DNA and enzymes have been detected in different isolates and a pattern has emerged which suggests that there are different strains of *E. granulosus*. In Britain there appears to be both a sheep/dog and a horse/dog strain, which are morphologically different. The Swiss cattle/dog strain also appears to be a distinct strain. Not all strains are infective to humans.

Knowing that strain differences do occur may have important consequences from the point of chemotherapy, potential vaccine development and serodiagnosis.

■ BOX 7.6 DISTRIBUTION

Hydatid disease or hydatidosis is a zoonotic disease which occurs thoughout the world, especially among communities where sheep and goat rearing is an occupation, using dogs to herd the flocks. This is reflected in differences in the disease occurrence between urban and rural populations. In Uruguay between 1962 and 1971 the incidence of hydatidosis in the urban populations was 10 cases for every 100,000 persons; however, in the rural areas it was 123 per 100,000. Similarly in Argentina the urban incidence of the disease was 2 per 100,000; and in the rural areas 150 per 100,000.

Another very endemic area is the Turkana region of Northern Kenya, with 15,000 diagnosed cases in a population of 143,000. The reason for the high incidence in that area is partly due to the fact that humans are not buried below ground and the villagers have a close association with domestic dogs.

■ SUMMARY

The presence and survival of a parasite can be the cause of pathology to the host. The multiplication of a protozoan within a cell and the subsequent release of the offspring which then re-invade new cells leads to cycle of continuous cell damage. In the gut mucosa this can bring about a change in the gut architecture and lead to a malfunctioning condition. Migration of the larger parasites through tissue and organs leads to permanent scarring of the tissues.

Once a parasite has established itself within the tissue, an innate inflammatory reaction follows. This could lead to either the destruction of the parasite or encapsulation within fibrous tissue. The permanent presence of a parasite can lead to destruction and atrophy of the surrounding tissue and malfunctioning of either the local tissues or the organ.

The trematodes that are the cause of a great deal of human disease are *Paragonimus*, *Opisthorchis* and *Schistosoma*. They inhabit the lungs, the liver and both mesentric and bladder blood vessels, respectively. Humans are infected via their direct or indirect contact with water. *Fasciola* is a trematode that mostly infects domestic animals, in particular sheep. There are cases of humans living in sheep-rearing areas becoming infected with *Fasciola*. The metacestodes of *Echinococcus* develop into large hydatid cysts and the disease that causes is known as hydatidosis.

END OF CHAPTER QUESTIONS

PATHOLOGICAL EFFECT OF THE PARASITE UPON THE HOST

Question 7.1 What main products produced by the parasite's metabolism can become toxic to the host?

Question 7.2 What effects can gut parasites have upon the host tissues?

Question 7.3 What chemical and physiological changes can occur as result of a parasitic infection?

Question 7.4 What are prostaglandins and leukotrienes and what is their function?

Question 7.5 What are the main characteristics of inflammation?

Question 7.6 Describe all the reactions that are part of the process of inflammation.

Question 7.7 What is a pyrogen and what are the functions of pyrogens?

Question 7.8 What interleukins are associated with inflammation?

REACTIONS TO PARASITIC HELMINTHS

Question 7.1 What cells are normally associated with reactions to parasites?

Question 7.2 Which cells usually are the first to 'attack' a parasitic infection?

Question 7.3 How do phagocytic cells destroy non-self material?

Question 7.4 Describe the sequence of events that eventually leads to the cellular destruction of a helminth parasite.

Question 7.5 Describe what happens to liver tissue when invaded by parasites.

Question 7.6 What is neurocysticercosis?

Question 7.7 Describe the changes that occur within the spleen.

Question 7.8 Discuss the various pathological conditions that can be caused by parasites.

HUMAN TREMATODE PATHOGENS

Question 7.1 Describe the life-history of three trematodes that infect humans.

Question 7.2 What diseases do these parasites cause?

Question 7.3 Where geographically are these diseases most frequently encountered?

Question 7.4 Describe the distribution of one of these infections in a particular endemic region.

Question 7.5 Name the different species of *Schistosoma* that can infect humans.

Question 7.6 What host avoidance mechanisms are usually associated with schistosome infections?

Question 7.7 Describe the pathology associated with schistosomiasis.

FASCIOLA HEPATICA AND FASCIOLIASIS

Question 7.1 Describe the life-history of *F. hepatica.*

Question 7.2 What are the main factors associated with the distribution of fascioliasis?

Question 7.3 Describe the disease fascioliasis.

Question 7.4 How do humans become infected with *F. hepatica* and what effect does the disease have upon them?

BIOLOGY OF *ECHINOCOCCUS*

Question 7.1 Name the two dominant species of *Echinococcus.*

Question 7.2 What is the difference between alveolar and hydatid disease?

Question 7.3 Describe the pathology caused by alveolar echinoccocosis.

Question 7.4 What disease does *E. granulosus* cause?

Question 7.5 Describe how the infection develops.

Question 7.6 What organs are usually affected?

Question 7.7 Describe the damage to organs that the disease can cause.

Question 7.8 In which parts of the world does the disease occur?

8 EPIDEMIOLOGY

Epidemiology is the study of the distribution of diseases in a particular community or district. A variety as well as a large quantity of data needs to be gathered in order to try and establish a pattern of distribution of parasitic infections. The data are best obtained either directly in the field or from a source such as hospital or regional health clinic records.

8.1 COLLECTION OF DATA

The type of data that have to be collected and collated when studying the distribution of a parasitic infection is based on the following:

- The total population of a defined geographical area and distribution of population within the area.
- Climatic factors, such as average temperatures, yearly rainfall, seasonal distribution of rain and temperature, humidity etc.
- The local topography, such as soil types, vegetation, water distribution (rivers, dams, lakes etc).
- The type of agricultural practices, soil husbandry, animal husbandry, fishing, forestry etc.
- Non-agricultural occupations and the amount of travel involved.
- Housing, fresh water storage and supply, sanitation.
- Diet, origin of food supply, storage, seasonal changes etc.

Population data should include age and sex profiles of the community and provide continuity of the population from one generation to another. The medical history of the population is important, if any particular disease is prevalent throughout the district.

Ecological data should include the type of fauna, invertebrates — insects snails etc and vertebrates, including domestic pets and agricultural animals.

Most epidemiological studies begin by acquiring data from the local health centre, clinic or hospital. This only provides information with regard to those persons who are already ill and seeking medical assistance — persons who are acutely or chronically ill. Such information helps to provide data about the true distribution or prevalence of a particular

■ BOX 8.1

The total number of patients examined over 29 months was just over 4,000 and the majority (72%) of these patients lived within a 5 km strip alongside a river.

- The mean annual temperature was 35°C with a range from +40°C to −4°C with no frost.
- The annual rainfall was 600 mm, of which 64% fell between January to April.
- The following types of parasite were detected by stool examinations of 518 patients (12.4% of the total examined):

Ancylostoma braziliensis and *A. caninum* (6.8%); *Strongyloides stercoralis* (2.7%); *Hymenolepis nana* (1%); *Schistosoma mansoni* (1%); *Taenia* spp (0.9%).

disease. In addition, identifying one particular parasite within an individual does not indicate whether or not he/she has any other parasitic infections.

Most studies of parasitic infections are distributed in the following way: that is, a few people have many parasites while most people have none or hardly any parasites. This suggests that some people have a degree of immunity to certain infections.

Methods of examination must be taken into account. For instance if only stools are examined, this will only reveal the presence of certain groups of parasites. It will not reveal the presence of the metacestode stage of cestodes nor filarial nematodes. Blood samples will show parasites associated with the circulating blood. Immunodiagnosis reveals the presence of tissue or organ parasites, but is not very specific. An important aspect of any epidemiological study is to ascertain whether or not the parasites found are zoonotic and if they identify the animal hosts.

Poverty often goes together with poor housing, lack of clean fresh water, inadequate sanitation, crowded living conditions and living in close proximity to domestic animals. These conditions are often associated with disease and parasitic infections. Wet, muddy and stagnant waters are frequently breeding grounds for vectors of infections.

A convenient method investigating the impact of parasitic infections in a community is to examine the patient records from a local hospital or health clinic. Many such studies have been carried out and reported and provide some interesting information with regard to parasite distribution in relation to the local topography and life styles etc. Such a study was a survey of patients from the rural hospital of an isolated community in Northern Namibia (Evans and Joubert 1989) (see Box 8.1).

An important finding was that many of those patients had more than one parasite, a condition referred to as polyparisitism. The age and sex of the infected individuals showed certain trends such as: More females were infected with hookworms — females while doing the family washing tend spend more time on damp ground where the infectious larvae live. *S. mansoni* was found mainly in males — males spend more in contact with water while fishing.

■ 8.1.1 RURAL COMMUNITIES

Surveys of rural communities in a variety of geographical areas have revealed that individuals are frequently infected with more than one species of parasite. This condition was clearly demonstrated in a series of investigations carried out by Buck et al. (1978) in the following areas: Chad and Zaire in central Africa, areas containing both dry savannah and humid wooded savannah; the low Amazon basin of Peru; and the semi-desert region of Afghanistan.

The areas chosen had similar-sized populations, with similar sex ratios and age distributions. The surveyors often encountered persons with multiple infections of closely related species. This condition can lead to difficulties in both treatment and diagnosis. For example in Zaire they found that five different species of filarial worms are coendemic and are often found within the same individuals.

A cross-sectional study of stools, urine, skin biopsies and blood samples showed that only a few persons at any age were entirely free of parasites. The majority had only one or two parasites, although a decreasing number were shown to be more heavily infected. In Laka tribesmen from central Africa, two or more parasites were detected in 38% of the population but only 3% (of the males) had at least four parasites.

The most frequent parasites were found in the following order: *Onchocerca*, *Schistosoma*, *Dipetalonema persistans* and malarial parasites.

Loa loa causes loaiasis; and its prevalence in Chaillu (central Africa) was found to be related to the ecology of the area. The humidity within the forests encouraged the breeding of the arthropod intermediate host. The infection rate in one village was found to be 35% and in another village, closer to the forest, it was found to be over 40%. Women and men were equally exposed, yet males over 30 years were more infected than females. Pygmy groups who were equally exposed to the infection in the same environment had significantly lower rates of infection.

Ascaris lumbricoides and *Trichuris* spp surveyed in children from slum areas in Malaysia showed that the prevalence was related directly to sanitation and population densities.

For a detailed study of an infection, such as visceral leishmaniasis in a forest area, to provide a complete picture of the spread and distribution of the infection it is important to obtain the following type of data:

- Social status, age, sex, ethnic group, occupation and reason for entering the forest.
- Clinical data: number of, localisation of and type of lesions.
- Biological data: mean diameters of the amastigotes on the smears.
- Date of appearance of the first lesion and the geographical site where the infection occurred.

Results from such surveys have shown that in all cases infections resulted from travelling into the forest either for work or leisure. No infections were observed in people who had never entered the forest. The legs and forearms, the most exposed parts of the body, were the most infected. The infection appeared seasonally, corresponding to periods of low rainfall — the peak time for the vector.

An important aspect of any epidemiological study is to ascertain whether or not the parasites found are zoonotic and if the animal hosts can be identified.

■ 8.1.2 RURAL–URBAN MOVEMENT

Studies of people who travel from an urban into a rural environment show an interesting distribution of parasites transmitted by arthropod vectors such as *Plasmodium* species. Blood samples taken from children in Lusaka, the capital city of Zambia, who regularly travelled into the surrounding rural areas produced some interesting results (Watts et al. 1990). Out of 423 urban children examined for blood parasites and serum antibodies, 10.3% were found to have live parasites and 62% had anti-malarial antibodies but no parasites. The presence of the antibodies was shown to be associated with journeys outside the towns, which suggests that they had suffered some form of malaria. Enlarged spleen (splenomegaly) is caused by an infection and the rate of splenomegaly in children

■ BOX 8.2

The last statement implies that host hormones might play an important role in development of immunity to *S. mansoni*:

- The hormones might have a direct influence on the parasite's metabolism.
- Induction in the host of physiological and anatomical changes such as increase in skin thickness, fat deposition etc, which affect the innate resistance.
- Changes to the immune system.

With regard to the immune system the following observations have been made:

- A steady increase in serum IgE antibody responses into adulthood.
- Serum levels of the antibody IgG_4 against adult worms is greatest at peak levels of infection.
- Anti-soluble egg antigen IgG_1 antibody levels in the serum are highest before the infection begins to decline.

from Lusaka was 3% in the urban areas, but 27% in the rural areas; and over 97% of these children were found to have anti-malarial antibodies.

Such surveys with regard to infection rates in a community indicate who and which people are likely to be susceptible and who will require prophylactic treatment. These values suggest that there was or is mesendemic malaria among rural children most of whom developed resistance to malaria. The investigations also indicated that although there is a low level of transmission in the urban areas, the urban children are more susceptible when they enter the rural areas because they have not developed any resistance.

Many of the parasitic helminth infections rise steadily to a peak in early teenage years and then fall off to relatively low levels in adulthood. A recent survey of *Schisotosoma mansoni* infections in Senegal (Fulford et al. 1998) has demonstrated the existence of such an age-intensity profile. Other studies in endemic areas have confirmed this fall-off of the infection rate during puberty (see Box 8.2).

■ 8.2 PROBLEMS AND DIFFICULTIES

It is important to understand and appreciate the problems and difficulties involved in any epidemiological study. In 1991 a survey of a *Schistosoma japonicum* infection was carried out among migrant fisherman in Dongting Lake in China (Li and Yu 1991). The population and geographical details are shown in Box 8.3.

For purposes of the investigations five sites were chosen; two each in the eastern and southern lakes and one in the western lake; and 652 boats with 1,842 persons were registered (details of each family were recorded). The boats were numbered and those used in the study were chosen at random. Eventually 122 boats with 316 persons over 3 years old formed the study group. This constituted 18.7% of the boats and 17.2% of the population.

The detection and treatment of patients, snail control, provision of water and sanitation and health education were carried out by the personnel at each field station. A questionnaire was used to determine the following data for each individual in the survey: sex,

■ BOX 8.3

There are series of fresh water lakes in northern Huna province covering an area of 15,200 sq. km. A series of small lakes plus rivers drains the high ground and flood waters rise up to 8 m before draining into the Yangtse river. One lake, the Dongting lake, has a surface area of 2,691 sq. km. It is the second largest fresh water lake in China and is one of the most endemic areas for *S. japonicum* in China. Malaria is not endemic in that region.

There are about 150,000 fisherman and families living in the area and most have extensive contact with the water. In general there are no modern sanitary facilities; and all urine and excreta enter the lake. The fisherman have no fixed homes and move around the lake according to the season and water level.

age and medical history such as: schistosomiasis symptoms; abdominal pain; diarrhoea (more than three bowel movements per day); bloody stools; weakness; absence from work due to illness; history of treatment of schistosomiasis.

Each person in the sample was given a physical examination to determine liver and spleen size. Faecal samples were collected and examined using the Kato–Katz technique. The parasite eggs per gram stool was calculated from three preparations. All positive cases were treated with a drug (praziquantel). Egg-positive cases were divided into three groups using a table of random numbers and adjusted for age and intensity of infection. Groups A and B were given a single oral dose of 40 mg praziquantel/kg body weight. Group C had no treatment at all. Group B was given further treatment after 6 months.

The data collected from this population showed that the overall infection was similar on all five sites. Among those infected there was a significantly higher prevalence in the males 49% (95/195) over females 28.9% (35/121), $P<0.01$. Two infected groups were detected between the ages of 20–29 (54%) and 40–49 (50%). The mean number of eggs per gram was 34.4 in the lower age group while in the higher group it was found to be 45.8 eggs per gram. Enlargement of the spleen (hepatomegaly) was diagnosed in 37.3% (118 subjects) and seven were found to have a palpable spleen. However the examinations indicated that enlargement of the liver and spleen was not significantly associated with egg secretion.

The effect of drug treatment (chemotherapy) was that 12 months post-treatment 77.8% and 84.1% of individuals in groups A and B respectively (the treated groups) no longer secreted eggs, whereas in the control group C only 14% were still not secreting eggs.

The above observations must be considered in conjunction with the local topography and climate. The lake levels rise up to 8 m in May to October. During this time faeces pass into the lake untreated. Infected snails are found throughout the year but are more numerous during the main fishing season; and during this period acute schistosomiasis is more common.

The migrant fisherman themselves were thought to be the most important source of the infection. However the possibility of a zoonosis, that is an animal reservoir host existing in the area, was investigated. Ten to fifteen per cent of the local water buffalo were found to be infected with *S. japonicum*. This is a much lower rate of infection than in the human population, but the volume and quantity of excreta from individual animals is so much larger than from humans that they may be an important source of contamination.

From such studies various ways of controlling the infection can be considered, such as reducing the level of faecal contamination and treatment of the fisherman, health education and wearing protective clothing and timing the treatment to coincide with the rise and fall of the water levels.

■ SUMMARY

Epidemiology is the study of the distribution of diseases in a given environment. A complete survey requires background data such as the local topography, climate, population size, sex ratio, age distribution, housing and occupations. Data are often obtained from local hospitals and health centres which may represent a true survey of all the local diseases. Examination of faecal, blood and urine samples are the most common methods of diagnosis. Once an overall view is gained of the parasites present and their distribution, methods of treatment and control have to be considered.

END OF CHAPTER QUESTIONS

Question 8.1 What is meant by epidemiology?

Question 8.2 What type of data is necessary for the study of the distribution of a parasitic infection?

Question 8.3 Where do most data collections begin and why?

Question 8.4 What type of diagnostic techniques is used to detect parasitic infections?

Question 8.5 Using actual examples explain how the local topography may account for the types of parasitic infections.

Question 8.6 What is meant by polyparasitism?

Question 8.7 Describe some examples of incidences of polyparasitism.

Question 8.8 Give examples how migratory behaviour can result in becoming infected with parasites.

Question 8.9 What results were obtained by a study of schistosome infections in Senegal?

Question 8.10 Discuss the findings of a study of a community where there is endemic schistosomiasis.

9 VACCINES

■ 9.1 THE IDEAL VACCINE

Until quite recently the only commercially available anti-parasite vaccine was Dictal (developed in 1959). Dictal is a vaccine against *Dictyocaulus viviparus*, the cattle lung worm and the cause of cattle bronchitis. A similar type of vaccine was developed in 1978 as protection against the dog hookworm *Ancylostoma caninum* but was only used in the USA.

The above two vaccines contain live, attenuated, irradiated larvae in capsules. More recently a vaccine against the cysticercus of the sheep tapeworm *Taenia ovis* has been manufactured using recombinant technology.

Serious attempts to produce an anti-parasite vaccine against the various human and domestic animal parasites have been ongoing since before the second world war. The great target for all would-be producers is to manufacture an anti-malaria followed by an anti-*Leishmania*, anti-*Trypanosoma* and an anti-*Schistosoma* (both human and cattle) vaccine. In addition there has been and still is research into protecting domestic animals against a variety of protozoan and helminth parasites, such as *Theileria*, *Trypanosoma*, *Eimeria*, *Fasciola hepatica*, *Schistosoma bovis* etc. Protection of cattle against *Schistosoma bovis*, *Theileria parva* and *T. annulata* has involved the use of live, attenuated infectious stages. Killed *Leishmania mexicana* together with BCG (Bacillus Calmette–Guérin) has been used in human trials and gave enough protection at the time to warrant further development.

An ideal vaccine will have to have the following properties:

- Generate enough antibodies to be able to recognise the presence of the parasite antigen.
- Prevent the parasite becoming established in its preferred site within the host's body.
- Allow for the boosting of antibodies during subsequent infections.
- Activate antibody-dependent cell-mediated cytotoxicity (ADCC).
- Induce a T cell cytotoxic-dependent response capable of killing the parasites directly or indirectly.
- Prevent the parasite from reaching any form of maturity and/or multiplying asexually.

■ 9.1.2 THE TARGET ANTIGEN

A vaccine has to identify that part of the parasite which constitutes the reactive antigen, the immunogenic antigen or the epitope, provoking an immune response. The epitope is usually part of the outer membrane (as in most protozoa) or the tegument (as in the trematodes and cestodes) or the cuticle (as in the nematodes). Secretions from the parasites in the form of metabolic waste products or 'enzymes' are released to assist with penetration and migration through tissues.

Once the epitope has been identified it has to be isolated, its molecular structure ascertained and then synthesised. Recombination technology is now used to produce synthetic antigens. After the compound proposed as a potential vaccine has been isolated, synthesised and purified, a route of administration has to be determined. The two most common methods are oral and/or through the skin.

The objective in producing an immunogenic antigen is to sensitise the host to a particular pathogen. The antigen stimulates the disease conditions but at much reduced or attenuated level. The adaptive immune system is stimulated and the eventual outcome is that the body will now have circulating primed B memory and T memory cells.

■ 9.2 THE DEVELOPMENT OF A POTENTIAL VACCINE

The essential ingredient to initiate the development of an anti-parasite vaccine is a plentiful supply of live parasite material in order to isolate the immunogenic antigen. If it is possible the target infection, ie the parasite, should be established in a laboratory host. The ideal situation would be to maintain the complete life-cycle of the parasite in a laboratory host/hosts, ie an *in vivo* culture of the parasite. If the above is not possible there are alternatives such as maintaining one stage of the life-cycle in vivo, using a laboratory host, and the other stages or phases of the life-cycle maintained in an in vitro culture. The asexual multiplicative phases of both protozoan and helminth parasites can often be adapted to surviving in vitro. A most important aspect is to maintain the stage of the parasite that is infective to mammalian hosts.

One of the many problems in the manufacture of a vaccine is the fact that most parasites have more than one antigenic identity. In addition some are capable of changing their antigenicity in response to the host's immune reaction.

A vaccine must be able to prevent the parasite from becoming established within the host. In order to achieve that objective, it is imperative to try and understand the host's normal/natural immune response.

One the most studied helminth parasites, with particular emphasis on immunity, is the digenean trematode *Schistosoma mansoni*. The complete life-cycle can maintained under laboratory conditions. The snail intermediate host *Biomphalaria glabrata* survives well in aquaria and the adult stages live for relatively long periods in most strains of laboratory mice. The *S. mansoni* model infection has demonstrated that certain degrees of immunity can be achieved by the presence of a living infection. The live parasite is protected against the immune response which it has stimulated by acquiring host proteins onto its tegument. These host proteins confuse the immune system into treating the parasite as host material. In other words acquired molecules which disguise the parasites provide a form of molecular disguise. This either prevents a secondary infection while keeping the primary infection under control and/or modulates the effect of the parasite upon the host, ie it immunomodulates the disease. The latter can have the effect of reducing the pathological effect of the parasite (see section 7.5.5).

For a discussion on the development of an anti-malarial vaccine see section 10.6.

■ 9.2.1 ADJUVANTS

In order to be certain of stimulating the immune response when the potential immunogenic antigen is weak, an adjuvant is used. An adjuvant is a substance that non-specifically enhances the immune response and is administered in combination with the vaccine. Adjuvants are often used when the antigen has a low immunogenicity or the amount of antigen available is restricted.

Adjuvants appear to act in the following ways: by prolonging the antigen within the host; enhancing the co-stimulatory signals; and finally stimulating non-specific lymphocyte proliferation.

The following compounds are among the most commonly used adjuvants:

- Aluminium potassium sulphate (alum), a salt that when mixed with the antigen precipitates the antigen and results in a slower release of antigen. This effectively increases the size of the antigen.
- Freund's incomplete adjuvant is made up of antigen in aqueous solution, with mineral oil and an emulsifying agent. The oil surrounding the antigen is dispersed in small quantities (droplets) and the antigen is released slowly from the site of inoculation.
- Freund's complete adjuvant in addition to the water-oil mixture contains a heat-killed bacterium *Mycobacteria*. The mycobacterial cell wall plus the muramyl dipetide component of the mycobacterial cell wall activates phagocytic cells and is more potent than the incomplete Freund's adjuvant.
- An established preparation often used an adjuvant is BCG (Bacille Calmette–Guérin), widely used as an immunising agent against tuberculosis. BCG contains an attenuated strain of *Mycobacterium bovis* and is also used an immunopotentiator with *Mycobacterium leprae* as an anti-leprosy vaccine.

In the process of developing specific acquired immunity to mycobacterial antigens, BCG vaccination induces an altered immune reactivity to heterologous antigens and hence is also widely used under experimental conditions as an adjuvant.

The immunity that is achieved in vaccinated subjects by using BCG is due to a specific anti-mycobacterial cell-mediated immune (CMI) response. The cell-mediated immune response after the BCG injection is due to the following affects: the activation of macrophages mediated by $CD4^+$ T cells; the subsequent proliferation of $CD8^+$ T cells; and the development of a granuloma.

The effects of using an adjuvant are an increased non-specific resistance against unrelated pathogens. Potentiated humoral and CMI responses to antigens are associated with or conjugated to non- or poorly immunogenic molecules. BCG has been shown to stimulate DNA synthesis in draining lymphoid tissues.

9.2.2 FACTORS THAT INFLUENCE THE DEVELOPMENT OF IMMUNE RESPONSES

Factors that influence the development and persistence of the specific and non-specific immune responses after BCG vaccination include the following:

- BCG substrains; in in vitro cultures of bacteria, new strains or varieties of the original bacteria may arise due to slight changes in the mediator growth conditions. These new colonies of bacteria may have slightly different physiological and biochemical properties which may render them pathological to the host.
- The inoculating dose of the living bacteria may also be critical, because an excessive amount of bacterial toxins may introduced.
- To introduce the bacterial antigen into the host, a mixture of dead and living organisms is inoculated at the site of vaccination. If the inoculum contains a larger propor-

tion of living bacteria, the host's immune system may not be able to contain them and they could then spread round the body. That is instead of remaining localised they become systemic.

- The usual route of administration is by injecting intra-muscularly. The major site or response is then localised but the subsequent reactive cells and antibodies need to circulate round the body to be protective. An oral route of infection may be more effective, but only after extensive testing can the correct route be determined.
- Once the bacteria in the vaccine have stimulated an immune response they should then be destroyed by the reaction that they provoked. Those bacteria that remain live could become virulent and potentially harmful.
- The host must be healthy and able to respond to the vaccine. If a host has not developed an immune system or its immune system is damaged or compromised then the vaccine will not be effective.

9.3 CUTANEOUS LEISHMANIASIS

A strain of mice known as BALB/c mice which are normally highly susceptible to *Leishmania* infection were treated with BCG and the result was that the severity of the infection was reduced. Under experimental conditions, naive hosts infected with *L. major* and viable BCG were found to have smaller lesions. BALB/c mice treated in a similar manner were protected against systemic infection. Complete protection against *L. major* cutaneous disease was achieved when BCG immune mice were challenged with viable BCG and parasites.

BCG has also been used in attempts to protect humans. A vaccine containing 6.4×18^{8} heat-inactivated *Leishmania mexicana* amastigotes plus BCG was prepared and administered to a group of volunteers. After three intra-dermal injections of the vaccine, 49 out of 52 sufferers of localised cutaneous leishmaniasis caused by the parasites of *L. brasiliensis* were cured within 32 weeks. A CMI response to *Leishmania* was instrumental in the recovery but whether or not the cure was due to the reacting antigen (the epitopes) of the *Mycobacteria* and *Leishmania* cross-reacting, or due to the cytokines produced modulating the immune response is matter for debate.

A vaccine that expresses *Leishmania* protective antigen in BCG is a possible goal. However if that target is to be achieved then there will have to be identification and cloning of the parasite genes responsible for the production of immunogenic antigens. Following that, there must be insertion of these genes into a bacterium so they can produce the antigens in vitro. The synthetic antigens have then to stimulate the host's response.

The current thinking with regard to the function of an anti-protozoan parasite vaccine is to sensitise the host so as to be able to prevent or control growth and multiplication of the parasite and reduce the pathological effect upon the host.

SUMMARY

Vaccination against the cattle lung worm (*Dictyocaulus viviparus*) is the only truly anti-parasite vaccine available. A vaccine should be able to produce memory cells in the host so that once contact is made with the pathogen, a rapid response to destroy the infection is set in operation.

The target antigen has to be clearly identified and produced either in culture or by recombination techniques so that, once admistered, it can stage an immune response without causing any of the disease symptoms. The response is an adaptive immune response, the memory of which is retained. The method of delivery and whether or not it requires an adjuvant is important. Anti-malaria, anti-leishmaniasis, and anti-schistosomais vaccines are currently being investigated.

END OF CHAPTER QUESTIONS

Question 9.1 Against which parasites have anti-parasitic vaccines been produced?

Question 9.2 What properties should an ideal vaccine possess?

Question 9.3 Describe a target antigen.

Question 9.4 Describe the methods that might be used in attempts to develop an anti-parasitic vaccine.

Question 9.5 What is an adjuvant?

Question 9.6 What are the essential properties of an adjuvant?

Question 9.7 What factors can influence an adjuvant?

Question 9.8 Describe actual examples of adjuvants that been used in attempts at anti-parasite vaccines.

ASPECTS OF MALARIA

10

The disease malaria is caused by the protozoan parasites belonging to genus *Plasmodium*. There are four species of malaria that infect humans: *P. falciparum*, *P. malariae*, *P. ovale* and *P. vivax*; and all are intracellular parasites. Apart from the initial multiplication stages in the hepatocytes, these parasites are entirely confined to infecting the erythrocytes in the circulating blood. To survive they replicate within the erythrocyte, the progeny escape from the red blood cell and each new individual invades a new erythrocyte. Each time this event occurs, parasite antigens in the form of metabolic waste products or secretions etc are released into the circulation. The major aspects of the epidemiology of malaria are:

- Children up to the age of six living in the rural areas in the tropics may have had malaria fever more than once. Older children often have the parasites in their system but do not have fever or any of the other symptoms of malaria: a condition referred to as 'clinical tolerance' or 'antitoxic immunity'.
- Many of the adults who live in endemic areas also have parasites in their blood but generally at a much lower level than observed in children.
- Adults rarely develop clinical malaria, although they continue to be infected from time to time. If immunity to malaria develops in such individuals it is acquired slowly despite continuous exposure to the infection, ie being continuously immunologically challenged. The immunity, if it occurs, is rarely complete.

10.1 OBSTACLES TO THE DEVELOPMENT OF A VACCINE

The main obstacles to the development of anti-malaria vaccine are:

The parasite has a unique habitat within the non-nucleated red blood cells (mature erythrocytes do not express HLA molecules). The parasite antigen may not be consistent from one generation to another, that is the antigen molecules are variable and may be presented in more than one shape (polymorphic).

Red blood cells (RBCs) infected with *P. falciparum* express the parasite antigens on their cell surface membrane, known as 'knob proteins'. These antigens expressed on the surface of infected RBCs bind to receptors on the endothelium and are considered to be a potential target for a vaccine against *P. falciparum*.

It is now established that antigenic variation occurs in *Plasmodium* infections but, unlike the African trypanosomes (*T. brucei*), the active region of the antigen is stable and could possibly be blocked by vaccines. The antigen and the receptors on the immune cells are closely linked. Phagocytic cells that take up the parasite antigens migrate to the lymph nodes and become antigen-presenting cells.

The molecular structure of the antigens of the different phases of *Plasmodium* have been analysed in detail. One of the characteristics of the antigens that has been determined is the fact that the major antigens are composed of repetitive sequences of amino-acids. Some of the features with regard to malaria immunity that have been established are:

- The repetitive sequences that dominate the structure of the antigens are presented to cells along T cell independent pathways.
- The parasite toxins which probably induce a cytokine response are thought to stimulate an innate non-specific (immune) mechanism which protects a non-immune person during the first few weeks of an infection.
- Tumour necrosis factor (TNF), a pyrogenic substance induced by the malaria toxins, appears to protect an individual against invasion by the parasites and at the same time induces a 'fever' which may itself be protective.

10.2 EFFECTOR MECHANISMS AGAINST THE SPOROZOITE INVASION

The first infective stage of the parasite, the sporozoite, is inoculated into the bloodstream by a mosquito feeding on the host's blood. The sporozoite exits from the circulating blood within 30 min and invades the liver parenchyma cells — the intrahepatocyte stage.

Any form of protection must be able to react rapidly to prevent invasion of the liver tissue. The antibodies and the activated immune cells (T cytotoxic cells) and the cytokines have to be present in sufficient concentrations for a complete and effective response.

The development of an anti-malaria vaccine must, like all vaccine production, centre around obtaining and analysing the immunogenic antigens. Antigen extracted from the sporozoite is known as the circumsporozoite antigen. The circumsporozoite protein isolated from in vitro cultures has the following amino-acid sequence: asparagine — alanine — asparagine — proline. These repeats are known as the $(NANP)_n$ sequence.

The synthetic production of the CS protein was the first attempt at producing a manufactured anti-malaria vaccine and has been extensively tested in laboratory infections using rodents, primates and human volunteers. Some of the findings of those experiments have shown that:

- Some groups of inbred strains of mice produce antibodies.
- In Gambian volunteers an increase in T cells in response to circumsporozoite protein was detected.

In attempts to produce a protective antigen, some of the peptides that comprise the circumsporozoite antigen have been synthesised to mimic the immunogenic molecules of the antigen (the epitope). These synthetic compounds were then injected into mice in order to develop a protective response. However, stimulation of the immune system using the synthetic 'antigens' did not always occur. This was probably due to the fact that the synthetic antigens did not associate with the MHC molecules on the macrophages and hence were not recognised by the antigen receptors on the T cells.

The use of experimental mice produced evidence that suggested that anti-sporozoite immunity can be cytokine-activated, or a cell-mediated activation, but not antibody-mediated. The vaccinated mice did not produce any detectable anti-CS antibody.

■ 10.3 HOST EFFECTOR MECHANISMS OPERATIVE AT THE HEPATIC STAGE

Soon after the initial inoculation into the blood stream, the sporozoites penetrate into hepatocytes (liver cells). The infected hepatocytes stimulate phagocytes in the liver to release cytokines and in particular gamma interferon (IFN-γ) The IFN-γ in the system activates infected hepatocytes to synthesise reactive nitrogen intermediates. The parasitised hepatocytes apparently also provide a signal for the generation of reactive nitrogen intermediates. A further cytokine, known as tumour necrosis factor alpha (TNF-α), is released from activated cells and induces the reactive nitrogen intermediates to kill infected cells. It is interesting to note that both IFN-γ and TNF-α are similarly needed to stimulate RNI-mediated killing of *Leishmania* by macrophages.

One of the first cytokines, interleukin-1 (IL-1), released by the activated macrophages stimulates the hepatocytes to produce the cytokine interleukin-6 (IL-6). This cytokine (IL-6) is thought to assist the reactive nitrogen intermediates as well as reactive oxygen intermediates to kill the intrahepatocytic parasites. However the parasites are still able to survive in this hostile environment apparently by down-regulating the production of IL-6.

■ 10.4 EFFECTOR MECHANISMS AGAINST ASEXUAL AND INTRAERYTHROCYTIC PARASITES

The majority of the in vivo studies of malaria used rodent models; in particular studies of *P. chabaudi*, which in many ways resembles human malaria. Severely compromised immunodeficient (SCID) mice were used in many of these studies.

The transfer of specific lymphocytes, first the T helper cells ($CD4^+$ T cells) and then antibody-producing cells (B cells), into SCID mice infected with the eyrthrocytic forms of *P. chabaudi* demonstrated that T helper cells ($CD4^+$) can be protective during the early stages of infection but the B cells are required to clear the infection.

Normal mice infected with *P. chabaudi* produce T helper cells (Th_1 cells) which secrete the cytokine gamma interferon (IFN-γ) and this prevents the infection from overwhelming the mouse. As the parasite becomes established within the mouse, the adaptive immune system is stimulated. After about 7 days, the activated B lymphocytes (B cells) give rise to antibody-secreting plasma cells and, during the later stages of the infection, the plasma cells secrete the antibody IgG_2 which helps to resolve the infection.

■ BOX 10.1

Severely compromised immunodeficient mice (SCID mice) are mice that lack most of the components of a normal immune system. Most of the missing components are lymphocytes. The mice are thus a very useful laboratory tool for studying how specific components of the immune system react to infection. In order to do so, immune cells taken from normal healthy mice of the same strain are injected into SCID mice. Hence single components or a combination of the components of the immune system can be investigated.

Further evidence of the importance of secreted cytokines in protecting the host is the detection of raised circulation levels of IFN-γ and TNF-α in humans infected mainly with *P. vivax* and *P. falciparum*. The fever suffered by patients with malaria may be due to the presence of elevated levels of the pyrogenic compound TNF-α. The increased body temperature may be inhibitory to the growth of the parasites.

■ 10.5 CEREBRAL MALARIA

Cerebral malaria is caused by *P. falciparum* and about 1% of *P. falciparum* infections cause childhood deaths. Erythrocytes containing maturing trophozoites become sequestered in the capillaries of brain, spleen, skeletal muscles, placenta and other internal organs. In the cerebral blood vessels the sequestration can be very dense. The exact mechanism causing the sequestration is still not clear and it is also not clear as to why only some children develop this condition. Patients suffering from cerebral malaria often relapse into a coma and what causes this condition is also not entirely understood.

The coma may be due to the cerebral blood vessels becoming blocked with infected erythrocytes attached or sequestered to the inner walls of the cerebral blood vessels. It was found that in some of the human clinical cases of cerebral malaria the sequestered cells in the cerebral blood vessels were exclusively parasitised erythrocytes, whereas in the mouse model the sequestered cells were found to be leukocytes.

Children with cerebral malaria were found to have significantly higher levels of circulating tumour necrosis factor alpha (TNF-α) than children infected with non-cerebral malaria.

In mouse cerebral malaria it was demonstrated that TNF-α can upregulate the intercellular adhesion (ICAM-1) molecules responsible for the adhesion of white blood cells to the walls of the blood vessels. An increase in the concentration of ICAM-1 molecules was detected in mice suffering from the mouse equivalent of cerebral malaria. In humans susceptibility to cerebral malaria may be due to whether or not TNF-α is activated.

It is not known why a child with cerebral malaria may be deeply comatosed one day and walking about the next. When parasitised erythrocytes in the cerebral blood vessels are sequestered, TNF-α is released locally and this stimulates the endothelium to release nitric oxide (NO). Nitric oxide is known to react with cells of nervous systems and disrupts transmission of stimuli between nerve cells. Some or all of the NO diffuses into the surrounding brain tissue and this could be the cause of the comatose condition.

Those patients who survive being infected with *P. falciparum* must have developed an immune response to the parasite to be able to control the infection. A prominent aspect of this response is the control of asexual multiplication of the merozoites, the stage that invades the eythrocytes. The monocytes, activated by toxins released by the parasite, secrete the pyrogenic compound TNF-α, ie this stimulates fever. The increase in body temperature may inhibit most of the growth of the parasites. The activated monoyctes and macrophages release cytokines that activate the T cells, both T helper and T cytotoxic cells. The Th_1 helper cells secrete IFN-γ which, together with TNF-α already released by activated macrophages, stimulates both the macrophages and neutrophils to attack the parasites.

In addition, the activated Th_2 helper cells act on the B cells which in turn produce antibody-secreting plasma cells. The antibodies are specific to *P. falciparum* antigen and may help to control the infection. The question that arises is: why is the infection not completely eliminated?

There is evidence to show that subsequent generations of the parasite (after asexual multiplication) have mechanisms to alter their antigenic configuration. The primary antibodies, which take about 7–10 days to be produced, are then no longer effective.

Once the concentration of IFN-γ increases it may inhibit the development of the Th_2 cells needed for the antibody response.

■ 10.6 ATTEMPTS TO DEVELOP ANTI-MALARIA VACCINES

The development and production of anti-malaria vaccine has been and still is the focus of much research. The primary development has been based mainly upon the use of in vitro and in vivo murine models. The problems associated with the development of an anti-malaria vaccine are due to the fact that the parasite has various phases in its life-history, such as the liver stage, the blood stage and the gametocytes. Apart from their physiological and biochemical differences, each phase has a different antigenic profile and the parasite appears to remain one step ahead of its host's complex immune system. The attempts to produce such a vaccine have followed two main strategies:

- To try and block the development of incoming parasites and prevent the development of the hepatic (pre-erythrocytic) and the blood (erythrocytic) stages.
- To limit the development of existing parasites.

However, recently two additional types of malaria vaccine have been proposed:

- A vaccine that would either neutralise the factors responsible for the pathology, ie an anti-disease vaccine.
- A vaccine that limits the transmission of malaria by immune interference with the parasite's life-cycle, particularly the sexual development in the mosquito.

■ 10.6.1 VACCINATION WITH IRRADIATED SPOROZOITES

It has been proposed and attempted to protect humans by immunising them with attenuated sporozoites delivered by the bite of irradiated mosquitoes. A group of volunteers were inoculated with sporozoites by multiple exposures to irradiated infected mosquitoes. Some individuals were apparently protected and were found to have high levels of anti-sporozoite antibodies in their circulating serum. Their peripheral T cells were found to respond to stimulation with a recombinant *P. falciparum* circumsporozoite (CS) protein and hence they were considered to be sensitised to the infection.

The production of such a vaccine is only feasible in laboratories that have access to mosquito-borne sporozoites. This problem can be partially solved by maintaining the sporozoites in an in vitro culture. The sporozoites are then inoculated artificially into primates and laboratory rodents, and such experiments have shown that antibodies are produced against the infective stage — the sporozoite.

The antigenic region of the sporozoite, that is the immunogenic molecules stimulating an immune response, is known as the circumsporozoite (CS) protein (see section 10.2). This antigenically active region (the epitope) contains proteins comprised of a series of repeat amino-acids known as CS repeats. If a synthetic version of these molecules could be vaccinated into the host, it could provoke an immune response that would then attack an inoculated sporozoite and prevent the parasites from becoming established within the host. The first attempts at preparing an anti-malaria vaccine incorporated the synthetic CS amino-acid repeats, and the initial human vaccination trials involved two preparations:

- A synthetic dodecapeptide $(Asn\text{-}Ala\text{-}Asn\text{-}Pro)_3$ conjugated to the tetanus toxoid. This represents the part of the epitope of *P. falciparum* CS protein.
- A non-conjugated recombinant polypeptide made up of a stretch of 32 *P. falciparum* CS repeat units $[(Asn\text{-}Ala\text{-}Asn\text{-}Pro)_{15}(Asn\text{-}Val\text{-}Asp\text{-}Pro)_2]$ fused to a stretch of 32 amino-acid residues.

Both vaccines only induced low levels of antibodies in most of the volunteers. To induce higher levels of antibodies, *P. falciparum* CS repeats were inserted into the hepatitis surface antigen (HB_s Ag) and the vaccine was adsorbed onto alum (an adjuvant) before being inoculated into volunteers. All of the volunteers developed antibodies to the *P. falciparum* CS repeats but the infection was still able to develop into the disease stage in many of the volunteers. A similar trial was carried out using *P. vivax*, another species of *Plasmodium* that infects humans, but the CS protein turned out to be poorly immunogenic in a trial involving 30 human volunteers.

10.6.2 CELL-MEDIATED IMMUNITY

Analysis of the immune response to malaria has shown that in addition to the production of antibodies the $CD8^+$ T cells (T cytotoxic cells) are activated. Cells not exposed to the malaria parasites can be extracted from body fluids from non-infected humans or mice and kept alive in culture cells. In culture, the cells can be exposed to specific circumsporozoite (CS) antigen extracted from cultured sporozoites. Non-infected mice were protected against strains of mouse malaria when injected with cytotoxic lymphocytes (CTLs) primed against the CS antigen. The target for the primed cytotoxic lymphocytes was found to be infected liver cells (hepatocytes). The cell adhesion molecules in the vicinity of the infected liver cells were activated and attracted primed CTLs, which led to the destruction of intrahepatocytic parasites. There is experimental evidence indicating that protection involved both cytotoxic lymphocytes and neutralising antibodies.

Vaccines that contain synthetic circumsporozoite peptides that may stimulate CTL activity to induce protection against malaria are now being prepared for trials.

10.6.3 BLOOD STAGE VACCINES

The parasites that invade the hepatocytes undergo a multiplicative phase, known as schizogony, which results in the production of numerous merozoites. The merozoites exit from the liver into the circulating blood and invade erythrocytes. Once inside an erythrocyte, the merozoite undergoes a development phase and grows into a trophozoite which has a very distinctive shape and is known as the 'signet ring' stage. Erythrocytes infected with this stage of *P. falciparum* can be maintained by in vitro culture and a parasite antigen, referred to as ring infected erythrocyte surface antigen (RESA) and identified as PF155/ring antigen, has been extracted from such cultures.

Infected erythrocytes have deposits of the antigen, a non-polymorphic 155 kDa polypeptide, on their surface membrane. The molecular structure of the antigen P155/RESA has at the carboxyl terminal region two tandemly repeated sequences (5 copies of Glu-Asn-Val-Glu-His-Asp-Ala and 30 copies of Glu-Glu-Asn-Val). These two regions of the antigen (the epitopes) react with receptor sites on human B and T cells.

Antibodies raised against the PF155/RESA antigens have been shown in vitro to be able to block the merozoites of *P. falciparum* from invading erythrocytes. Partial protection was achieved in monkeys by passive transfer of IgG antibodies specific to either the $(Glu\text{-}Glu\text{-}Asn\text{-}Val)_2$ or the Glu-Glu-Asn-Val-Glu-His-Asp-Ala repeats.

Trials are being conducted using vaccine containing synthetic PF155/RESA antigens.

10.6.4 MEROZOITE STAGE VACCINES

The 'signet ring' trophozoite stage develops within the erythrocytes into schizonts and, after they undergo the process of schizogony, they release blood stage merozoites capable of invading new red blood cells. Newly formed *P. falciparum* merozoites have been isolated and from these merozoites surface polypeptide antigens have been extracted, known as the merozoite surface antigen-1 (MSA-1) and merozoite surface antigen-2 (MSA-2).

Both MSA-1 and MSA-2, like the circumsporozoite proteins, have also been tested as a potential anti-malaria vaccine. *Aotus* monkeys (howler monkeys) vaccinated with purified *P. falciparum* MSA-1 like material induced protection against being infected under laboratory conditions with infected *P. falciparum* erythrocytes. Using the same type of technology, a similar vaccine from the merozoites of *P. vivax* has been prepared for testing.

This merozoite surface antigen MSA-2 varies in size (45 and 55 kDa) and contains a highly diverse central region of approximately 160 amino-acids. In order to use it as a vaccine it has to be conjugated to a *Diphtheria* toxoid. T helper cell responses in mice were induced by the MSA-2 proteins.

10.7 A SYNTHETIC VACCINE

In order to overcome the problems of having to isolate antigens from the various stages, a synthetic polymeric blood stage vaccine was produced, called SPF 66. It contains three short peptides representing sequences from three *P. falciparum* blood stage antigens and was prepared and tested in Colombia on volunteers. The vaccine was composed of peptides SPF 35.1 and SPF 55.1, based on partial sequences obtained from two as yet unidentified *P. falciparum* molecules. The first test results showed the SPF 66 vaccine to be immunogenic and safe to use in humans. A further set of trials was carried out on another 352 volunteers. The immunisation schedule was 2 mg doses of SPF 66 in alum at days 0, 30, 180. Unfortunately the results of the trials did not establish SPF 66 as a potential vaccine. The effects of SPF 66 vaccine on *P. falciparum* infections were that two of the peptides (in the vaccine) bound to erythrocytes and inhibited invasion of a number of erythrocytes.

Although there have been great advances in the understanding of the immune response to the malaria parasite, a vaccine to protect against being infected is still not available.

SUMMARY

There are four species of *Plasmodium* that are the cause of human malaria, which is one of the main causes of child deaths in large areas of the world. The intracellular habitat of *Plasmodium* helps to protect it from destruction by the host's immune response. Each phase of the parasite, ie the sporozoite, the merozoite and the gamont have a different antigenicity, with each inducing a host response. The effector mechanisms against all of the different stages have been studied in detail, mainly by the examination of serum from infected patients. There is an ongoing programme to try and produce a vaccine against *Plasmodium*, using all of the antigenic stages of the parasite.

END OF CHAPTER QUESTIONS

Question 10.1 What parasites are responsible for the disease malaria?

Question 10.2 What are possible causes of the symptoms associated with malaria?

Question 10.3 What are the main obstacles to developing an anti-malarial vaccine?

Question 10.4 Describe the effector mechanisms used against the sporozoite invasion.

Question 10.5 What are the host effector mechanisms used against the hepatic stage of the parasite?

Question 10.6 Discuss the effector mechanisms used against the intraerythrocytic stages?

Question 10.7 Which of the malaria parasites is the cause of cerebral malaria?

Question 10.8 What is the function of the CAM molecules with regard to malaria?

Question 10.9 What is the role of T cells in the reaction to the malaria parasites?

Question 10.10 What controls the erythrocytic asexual cycles?

Question 10.11 In the attempts to produce an anti-malaria vaccine, what are prime objectives of the vaccine?

Question 10.12 What different types of vaccine have tested?

Question 10.13 Name the different types of vaccines that have been produced by recombinant technology.

SUPPLEMENTARY READING

GENERAL

Cox, F.E.G. (ed.) (1996) *Modern Parasitology: A Textbook of Parasitology*. Published by Blackwell.

Katz, M., Despommier, D.D. & Gwandz, R. (1988) *Parasitic Diseases*, 2nd edition. Published by Springer.

Schmidt, G.D., Mayer, M.C. & Olsen, O.W. (1992) *Essentials of Parasitology*, 5th edition. Published by WCB.

Smyth, J.D. (1994) *Introduction to Animal Parasitology*, 3rd edition. Published by Cambridge University Press.

HOST RESPONSE

Benjamini, E., Sunshine, G. & Leskowitz, S. (1996) *Immunology: A Short Course*, 3rd edition. Published by Wiley-Liss.

Janeway, C.A. & Travers, P. (1997) *Immunobiology: The Immune System in Health and Disease*, 3rd edition. Published by Current Biology.

Kuby, J. (1997) *Immunology*, 3rd edition. Published by Freeman.

Roitt, I., Brostof, J. & Male, D. (1993) *Immunology*, 3rd edition. Published by Mosby.

PHYSIOLOGY AND NUTRITION

Arme, C. (1996) Nutrition. Chapter 5 in: Cox, F.E.G. (ed.) *Modern Parasitology: A Textbook of Parasitology*. Published by Blackwell.

Bryant, C. (1996) Biochemistry. Chapter 3 in: Cox, F.E.G. (ed.) *Modern Parasitology: A Textbook of Parasitology*. Published by Blackwell.

Chappell, L.H. (1996) Physiology. Chapter 4 in: Cox, F.E.G. (ed.) *Modern Parasitology: A Textbook of Parasitology*. Published by Blackwell.

Cornford, E.M. (1991) Regional modulations in tegumental glucose transporter kinetics in the rat tapeworm. *Exp Parasitol* 73:4 489–499.

Kawakami, Y., Kojima, H., Nakamura, K., Suzuki, M., Uchida, A., Murata, Y. & Tamai, Y. (1995) Monohexoylerceramides of larval and adult forms of the tapeworm *Spirometra erinacei*. *Lipids* 30:4 333–337.

Stradowski, M. (1995) Comparison of the viability of tapeworms obtained as as result of infecting rats with *Hymenolepis diminuta* larave of different ages. *Wiad Parazytol* 41:2 211–216.

Stradowski, M. (1995) Intensified morphological variability during patency in *Hymenolepis diminuta* originating from inbred tapeworm–WMS IL1. *Wiad Parazytol* 41:1 33–41.

Wang, C.C. (1984) Parasite enzymes as potential targets for antiparasitic chemotherapy. *J Med Chem* 27:1 1–9.

MALARIA AND VACCINES

Butcher, G.A. (1995) Sarcoidosis and malaria. *Immunol Today* 16: 252.

Patarryo, M.E. & Moreno, A. (1995) Malaria vaccines. *Curr Opin Immunol* 7: 607–611.

Riley, E. (1995) Malaria vaccine trials: SPf66 and all that. *Curr Opin Immunol* 7: 612–616.

SPECIFIC REFERENCES

Barnish, G. & Ashford, R.W. (1989) *Strongyloides cf. fuelleborni* and hookworm in Papua New Guinea: patterns of infection within the community. *Trans Roy Soc Trop Med Hyg* 83: 684–688.

Buck, A.A., Anderson, R.I. & MacRae, A.A. (1978) Epidemiology of poly-parasitism. I. Occurrence, frequency and distribution of multiple infections in rural communities in Chad, Peru, Afghanistan and Zaire. *Tropenmed Parasitol* 29: 61–70.

Chernin, J. & McLaren, Diane, J. (1983) The pathology induced in laboratory rats by the metacestodes of *Taenia crassiceps* and *Mesocestoides corti*. *Parasitology* 78 (3): 279–287.

Chernin, J., Miller, H.R.H., Newlands, J. & McLaren, D.J. (1988) Proteinase phenotypes and fixation properties of rat mast cells in parasitic lesions caused by *Mesocestoides corti*: selective and site specific recruitment of mast cell subsets. *Parasite Immunology* 19: 433–447.

Evans, A.C. & Joubert, J.J. (1989) Intestinal helminths of hospital patients in Kavango territory Nambia. *Trans Roy Soc Trop Med Hyg* 83: 681–683.

Fulford, A.J.C., Webster, M., Ouma, J.H., Kimani, G. & Dunne, D.W. (1998) Puberty and aged related changes in susceptibility to Schistosome infection. *Parasitol Today* 14:1 23–26.

Gottstein, B. & Felleisin, R. (1995) Protective immune mechanisms against the metacestode of *Echinococcus multilocularis*. *Parasitology Today* 11 (9): 320–326.

Hildreth, M.B., Pappas, P.W. & Oaks, J.A. (1997) Effects of tunicamycin on the uptake and incorporation of galactose in *Hymenolepis diminuta*. *J Parasitolol* 83:4 555–558.

Horuk, R. (1994) The interluekin-8-receptor family: from chemokines to malaria. *Immunol Today* 15: 169–174.

Ito, A., Honey, R.D., Scanlon, T., Lightowlers, M.W. & Rickard, M.D. (1988) Analysis of antibody responses to *Hymenolepis nana* infection by the enzyme-linked immunosorbent assay and immunoprecipitation. *Parasite Immunol* 10:3 265–267.

Kemp, M., Theander, T.G. & Kharazmi, A. (1996) The contrasting roles of $CD4^+$ T cells in intracellular infections in humans: leishmaniais as an example. *Immunol Today* 17: 13–16.

Li, Y.S. & Yu, D.B. (1991) *Schistosoma japonicum* infection among migrant fishermen in the Dongting Lake region of China. *Trans Roy Soc Trop Med Hyg* 85: 623–625.

McCracken, R.O. & Lipkowitz, K.B. (1990) Structure-activity relationships of benzothiazole and benzimidazole anthelminthics: a molecular modeling approach to in vivo drug efficacy. *J Parasitol* 76:6 853–864.

McCracken, R.O. & Taylor, D.D. (1983) Biochemical effects of thiabendazole and cambendazole on *Hymenolepis diminuta* (Cestoda) in vivo. *J Parasitol* 69: 295–301.

Mckeever, D.J. & Morrison, W.I. (1994) Immunity to a parasite that transforms T lymphocytes. *Curr Opin Immunol* 6: 564–567.

Medzhidtov, R. & Janeway, C.A. (1997) Innate immunity: impact on the adaptive immune response. *Curr Opin Immunol* 9: 4–9.

Pearce, E.J. & Reiner, S.L. (1995) Induction of Th2 responses in infectious diseases. *Curr Opin Immunol* 7: 497–504.

Reiner, S.L. & Seder, R.A. (1995) T helper cell differentiation in immune response. *Curr Opin Immunol* 7: 360–366.

Richter, J., Poggensee, G., Helling-Gieses, G., Kjetland, E., Chitsulo, L., Koumneda, N., Gundersen, S.G., Krantz, I. & Fekdmeier, H. (1995) Transabdominal ultrasound for the diagnosis of *Schistosoma haematobium. Trans Roy Soc Trop Med Hyg* 89: 500–507.

Stadecker, M.J. & Villaneuva, P.O.F. (1994) Accessory cell signals regulate Th-cell responses: from basic immunology to a model of helminthic disease. *Immunol Today* 15: 517–574.

Scharton-Kersten, T.M. & Sher, A. (1997) Role of natural killer cells in innate resistance to protozoan infections. *Curr Opin Immunol* 9: 44–51.

Stradowski, M. (1996) The effect of crowding tapeworms *Hymenolepis diminuta* base on proglottids in period of intensified vaiability. *Wiad Parazytol* 42:2 185–95.

Watts, T.E., Wray, J.R., Ng'andu, N.H. & Draper, C.C. (1990) Malaria in an urban and a rural area of Zambia. *Trans Roy Soc Trop Med Hyg* 84: 196–200.

INDEX